THE NATURE OF
FOXES

THE NATURE OF
FOXES
Hunters of the Shadows

REBECCA L. GRAMBO

GreyStone Books

Douglas & McIntyre

Vancouver/Toronto

Greystone Books
A division of Douglas & McIntyre
1615 Venables Street
Vancouver, British Columbia V5L 2H1

Published in the United States of America by Sierra Club Books, San Francisco.

CANADIAN CATALOGUING IN PUBLICATION DATA
Grambo, Rebecca L., 1963–
    The nature of foxes

    Includes bibliographical references and index.
    ISBN 1-55054-184-6

    1. Foxes. I. Title.
QL737.C22G72 1995   599.74'442   C95-910102-0

Editing by Jane McHughen and Nancy Flight
Front jacket photograph of red fox by Tom and Pat Leeson
Back jacket photograph by Alan and Sandy Carey
Jacket and text design by Barbara Hodgson
Printed and bound in Singapore by C.S. Graphics Pte. Ltd.

The publisher gratefully acknowledges the assistance of the Canada Council and the British Columbia Ministry of Tourism, Small Business and Culture for its publishing programs.

PAGE IV–V: *Like most foxes, this red fox needs to have a readily available source of water. Some desert-dwelling foxes, like the kit fox, have broken free of this restriction by getting the moisture they need from the food they consume.* TOM & PAT LEESON

PAGE VI: *Looking like the particularly cute creation of an imaginative toy maker, the diminutive fennec is actually a skilled predator that is highly adapted for life in the African desert. The smallest fox of all, it rarely weighs more than 1.5 kg (3 pounds).* GLEN & REBECCA GRAMBO

FOR MY MOTHER, DELORES,

WHO ALWAYS TOOK TIME TO

SHARE MY DISCOVERIES

CONTENTS

At rest near its den, a kit fox lifts its head to listen. Kit foxes, North America's version of the fennec, live in the arid regions of the western United States and northern Mexico. The very similar swift fox hunts nearby on North America's prairies. GLEN & REBECCA GRAMBO

PREFACE

There is a magic about foxes that is difficult to define. Words such as *graceful, beautiful* and *intelligent* don't seem to completely describe what makes foxes so fascinating. Perhaps the attraction comes from the fact that we rarely see foxes unless they wish us to do so. For me, there is always a feeling of privilege when I am allowed a glimpse into the life of these remarkable creatures. The many hours I have spent watching foxes and other animals are some of the most treasured times of my life.

The first live fox I ever touched was a hand-raised silver-phase red fox named Fancy. As my fingertips travelled over her face, I was stunned by the delicacy of the underlying bone. Beneath her thick coat I could feel the wiry strength of her surprisingly small body. Fancy thoroughly investigated my clothing, not forgetting to probe in each pocket and sample my shoelaces. She seemed quite taken with my camera and flash, both of which now display small toothmarks as tangible evidence of fox curiosity.

Another memorable fox encounter occurred while I was travelling home from a photo shoot with my husband, Glen, who was driving. I happened to glance out across a green pasture on the other side of the highway and caught a flash of red in the tall grass. I yelled at Glen to stop, and a few minutes and one or two near-accidents later, we were parked by the roadside watching a red fox vixen and her six kits romping in the late afternoon sun. One minute the kits were completely invisible, and the next, the air was filled with a swirling mass of small fox feet and tails. A particularly myopic ground squirrel ambled carelessly out into the pasture, but the foxes were so intent on their games that it escaped unscathed. We watched, marvelling at the exuberance of the kits and the patience of their mother, until the vixen finally led her babies off into the trees for the evening's hunt. Like many of my meetings with foxes, this was completely unexpected—a glorious bonus.

FACING PAGE: *Although most foxes do their hunting at night, some are active during the day and others prefer the shadowy hours of dawn and dusk, curtailing their travels during the extremes of darkness and light. Activity patterns vary with species and season, and even to some extent among individuals.* ERWIN & PEGGY BAUER

So many people have helped in various ways with this project. My heartfelt thanks to Candace Savage and Rob Sanders for their patience and encouragement. Jane McHughen offered much helpful advice, and Nancy Flight used her editorial skills to fine-tune the text. Ludwig N. Carbyn, research scientist with the Canadian Wildlife Service in Edmonton, Alberta, Canada, reviewed the manuscript, and David Macdonald, Jennifer Sheldon, Charles MacInnes, Gene Trapp and Laurie McGivern answered many questions. David and Cathryn Miller assisted with proofreading. Brad and Kathy Peters were kind enough to introduce me to Fancy and their other foxes.

Thanks to everyone who called with fox sightings and invited us to come stay while we fox-watched, especially Joan and Del Foulston, Clayton Cave, Doug and Darlene Thiessen, Sherry and Allen Aitken, Peter Law, Bill Meekins, Dennis Senholt and members of the Saskatchewan Wildlife Art Association.

My family supplied essential services. My sister, Jane, kept up a steady flow of supportive messages and brightened my days with a great variety of fox pictures. My husband, Glen, was unwavering in his interest and encouragement, listening to an unending stream of fox facts without complaint. Freddy, my large lop rabbit, provided comic relief and quiet companionship. Thank you all very much, and I sincerely promise a change of conversational topics.

For me, the magic of foxes has grown even stronger. I hope that this book brings you closer to an animal that many of us take for granted but know little about. As you learn more of their remarkable lives, perhaps you will share my fascination with foxes.

FACING PAGE: *Short legs churning at high speed, an arctic fox races through the snow. As it searches for food, those short legs carry the fox over huge distances; arctic foxes are second only to humans in longest terrestrial treks made by mammals.* ALAN & SANDY CAREY

Chapter 1 **FOXES AROUND THE WORLD**

As I stepped into the clearing, a splash of colour caught my eye. A red fox lay napping in the long grass. A breeze ruffled his fur, kindling sparks in the late afternoon sunlight, carrying my scent to him. One golden eye slowly opened and the fox watched as I sank to the ground. Eventually, with a catlike stretch and a tremendous yawn, the fox rose. He trotted towards the woods, zigzagging to investigate promising sounds and smells. At the clearing's edge he gave me one last glance, melted into the trees and was gone.

Through the lengthening evening shadows that mark the close of the day, foxes glide with silent grace. From dusk to dawn, they hunt through lush rain forests and on windswept ice. They pad across desert sands and over wet pavement streaked with the reflections of street lights. There are twenty-two species in all, with representatives on every continent except Antarctica. They may be small or large, light or dark, their form dictated by the demands of their specific environment. Yet all are instantly recognizable as "some kind of fox," with their pointed faces, intelligent eyes, alert ears, elegant whiskers and luxuriant, plumy tails.

Foxes belong to the dog family, Canidae, along with wolves, coyotes and your pet poodle. Yet many of their characteristics are distinctly feline. About forty million years ago, dogs and cats shared a common ancestor in the weasel-like Miacidae. After that, however, the feline and canine families diverged, with the fox sprouting from the canine family tree. The strong likeness between foxes and cats is a result of convergent evolution—similar lifestyles causing two unrelated groups to develop along parallel lines.

Foxes and cats both seek prey that is small, providing only enough food for a single hunter. The wary nature of their quarry demands the stealth of a solitary pursuer. When the fox senses prey nearby, it freezes, straining to locate a telltale scurrying in the grass and

FACING PAGE: *Resting but still alert, a red fox soaks up the warmth of the summer sun. Foxes around the world have to be on the lookout not only for prey but for predators too. Coyotes, jackals, leopards and eagles all kill foxes, but the biggest threat comes from humans.* GLEN & REBECCA GRAMBO

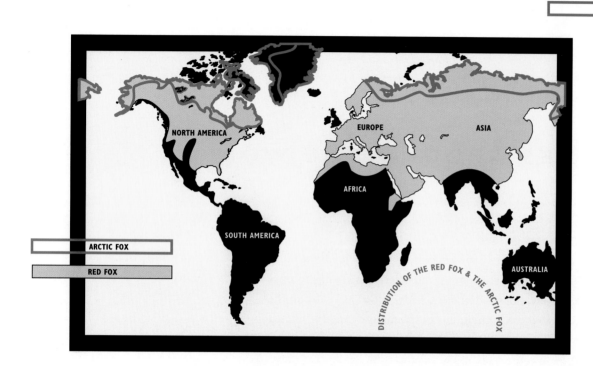

NORTH AMERICA · EUROPE · ASIA · AFRICA · SOUTH AMERICA · AUSTRALIA

ARCTIC FOX

RED FOX

DISTRIBUTION OF THE RED FOX & THE ARCTIC FOX

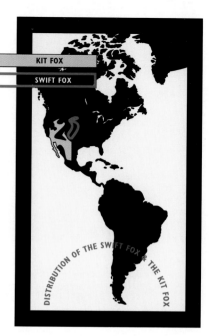

KIT FOX

SWIFT FOX

DISTRIBUTION OF THE SWIFT FOX & THE KIT FOX

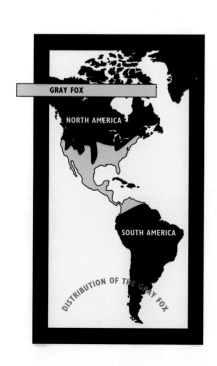

GRAY FOX

NORTH AMERICA

SOUTH AMERICA

DISTRIBUTION OF THE GRAY FOX

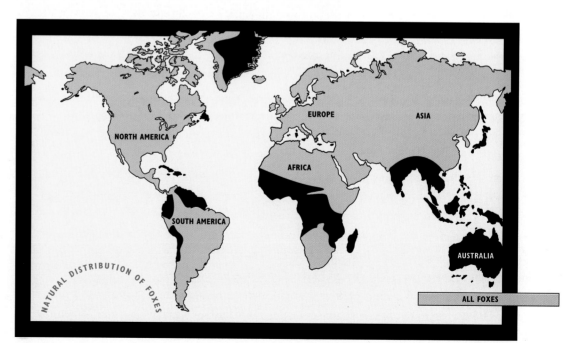

NORTH AMERICA · EUROPE · ASIA · AFRICA · SOUTH AMERICA · AUSTRALIA

ALL FOXES

NATURAL DISTRIBUTION OF FOXES

holding its body taut with concentration. With a twitch of its tail, it pounces, stabbing forward with nose and forepaws to pin its prey to the ground. Only then does the fox know what it has captured, having found its quarry by sound rather than sight. This "mousing leap" is behaviour with which any cat owner will be familiar.

When a quarry that demands a tactic other than the mousing leap captures their interest, foxes and cats switch gears to a slow, deliberate stalk. With their bellies held low to the ground, they creep forward, eyes fixed unwaveringly on their prey. Each paw is carefully lifted and advanced, hind foot placed precisely into the spot just vacated by a front foot.

A close look at fox or cat feet reveals soft, supple toe pads and tufts of hair between the toes, features that help to muffle the animal's footsteps. The red fox even has partly retractable catlike front claws. On their muzzles and wrists, most foxes and cats have groups of whiskers that are joined to special sets of nerve cells and are extremely sensitive to any contact. In cats, and probably in foxes, information gathered by the whiskers is transmitted to the brain in the same form as information from the eyes. In their low-light lifestyle, foxes and cats rely on this combination of sight and touch to build a clear image of their surroundings.

As Sandra Sinclair describes in *How Animals See*, animals that are active at night, like the fox and the cat, need very sensitive eyes—even when hunting in an open meadow under the brightest of full moons, a fox has millions of times less light to see by than it would during the day. Nocturnal animals have evolved a special layer, the tapetum, behind the retina in their eyes. The tapetum reflects incoming light back to the light-receptive parts of the eye, which are called photoreceptors, giving them a second chance at light that would otherwise be lost. To protect the sensitive photoreceptors from the abundant light during the day, the eyes of both cats and foxes have vertical slit pupils, which can shut more tightly than round pupils. Considering the dusk-to-dawn lifestyle of the fox and cat, the shape of their pupils makes perfect sense.

Foxes and cats do sacrifice something to have eyes that work well in dim light—image clarity. The eyes of foxes and cats, like our own, contain two kinds of photoreceptors, called rods and cones. The information from rods, which function in low light, is pooled before being transmitted to the brain, yielding a low-resolution image. Cones, which distinguish one colour from another, require more light to work. Unlike rods, individual cones send information to the brain separately, providing a sharply detailed image. Most animals active during low-light hours have eyes dominated by rods. They can detect movement better than we can in the dark, but what they see lacks colour and detail.

For a hunting fox, hearing is perhaps the most important sense. Tests show that a fox's hearing is most sensitive to low-frequency noises, the kind of rustling and gnawing noises made by small animals. The red fox can locate such noises to within centimetres of their

OPPOSITE: *Foxes naturally inhabit four of the six continents and were introduced to a fifth, Australia. Human extermination of competing predators, such as the wolf and coyote, helped one fox species, the red fox, become the most widely distributed carnivore in the world. Other foxes are adapted for life in a specific habitat. Arctic foxes hunt on the tundra, whereas small-eared dogs make their home in the lowland rain forests of South America. Several foxes, including the kit, fennec, Sechuran and pale fox, live in deserts. Of the twenty-two species of fox around the world, the red, arctic, swift, kit and gray are the ones we know most about.*

World distribution map: Based on D. W. Macdonald, 1984, The encyclopedia of mammals (New York: Facts on File).

Other maps: Based on M. W. Fox, 1975, The wild canids: Their systematics, behavioral ecology, and evolution (New York: Van Nostrand Reinhold).

Canadian researcher David Henry analyzed how the physiology of the red fox helps make its catlike mousing leap the highly efficient method of capturing prey seen here. First, compared with other canids, the red fox has light, narrow bones as well as a small stomach that reduces the amount of weight it can take on from a meal. By being lighter, the red fox can jump farther. Second, the hind legs of the red fox are disproportionately long compared with those of some of its close fox relatives. Longer legs maintain contact with the ground for a slightly greater period of time, allowing the fox to generate more thrust when pushing off for a lunge. Finally, Henry observed that the red fox takes off at an angle just slightly less than the 45 degrees needed for maximum distance. He theorizes that to minimize wind effects, the fox uses the lowest angle that will produce the needed distance. When a red fox wants to pounce on something close by, it simply increases its angle of takeoff, thereby reducing the length of its jump. Amazingly, all of this assessing and adjusting is done instinctively in the split second before the fox lunges, arcing forward like a lethal, beautiful missile. ERWIN & PEGGY BAUER

RIGHT: Ears that can pinpoint a telltale scurrying in the grass, whiskers that carry important tactile information to the brain, a nose capable of detecting very faint scents, and slit pupils to protect extremely sensitive eyes—in the fox, nature has fashioned a highly efficient low-light hunter. GLEN & REBECCA GRAMBO

true source. Even when resting a fox constantly moves its ears, like a radio listener scanning the available channels. In his book *Running with the Fox*, British wildlife researcher David Macdonald vividly describes the perceptions of a red fox:

> Much of the fox's life is spent on a knife-edge, deluged by the acuteness of its senses. In the fox, evolution has fashioned a creature for which every input is tuned to maximum sensitivity: for the fox there is the jolting image of the rabbit's blinking eyelid, the clamorous squeak of a mouse 20 meters off, the dreadful reek of a dog's day-old paw print.

Acute hearing, superb sense of smell, excellent night vision—you can see why death for a small mammal often comes softly on four black paws.

Small mammals are the staple item on most foxes' dinner plates, but the menu varies widely. When fruit is in season, foxes may eat it almost exclusively. If baby cottontail rabbits are in abundance, they become a major diet item. Some foxes eat more plant material than others, but almost all seem to include some type of vegetable matter in their diet. Then there are preferred items on each individual fox's list. Just as you might choose corn over broccoli, a fox would rather have a vole than a shrew. The red foxes where I live, for example, seem exceptionally fond of wild strawberries.

Whenever a fox has more food than it can immediately devour, it stockpiles the leftovers as insurance against future shortages. Most foxes show this frugal behaviour, which is known as caching. When the time comes to relocate their hidden resources, the animals use a combination of "public" and "private" methods. If the fox uses its visual memory ("under the oak tree with the big crooked branch"), it is employing a private method only available to the fox that made the cache. Locating the cache by smell is a very public method used not only by the fox that originally stored the food but by a variety of cache robbers, including bears, coyotes and other foxes.

Foxes tend to be solitary in their hunting and caching habits, but research has revealed a social side to fox life as they come together to rear a family each year. Whether a fox pair mates for life, as biologist David L. Chesemore reports arctic foxes may, or only for a season, they can be nearly inseparable while courting. The male follows his chosen female, spending time with her, gradually getting closer and touching her. This courtship period allows the two normally solitary animals to grow accustomed to each other. The culmination of this process comes in the winter during the few days when the vixen is receptive to breeding. Once mating has taken place, the male fox may not be as constantly attentive, but he is there to help the vixen when the kits are born about fifty to sixty days later.

During the courtship period, the female fox visits and cleans out several dens, eventually choosing one in which to give birth. Foxes don't usually dig their own dens, preferring to

FACING PAGE: *The large ears of the bat-eared fox help to gather and focus the noises of its insect prey and also aid in regulating its body temperature on the hot African savannah. Insects make up 80 per cent of the bat-eared fox's diet, and when it munches them, it does so with more teeth than any other canid.* ART WOLFE

use abandoned burrows of other animals such as badgers, ground squirrels and marmots. The vixen makes whatever changes are needed to convert the den to prime fox housing and settles down nearby to await the birth of her kits. After giving birth, she may be completely dependent on her mate for food. He hunts for both of them, bringing food to the den where she stays, caring for their new offspring. As the kits grow older, both parents share the heavy task of providing enough food for the family. Should one parent be killed, the survivor will often try to raise the kits on its own.

Fox living arrangements vary according to species and even within a single species. Some foxes live as single families, whereas others form larger groups of one male and several vixens. Those vixens are usually related to each other, as mother and daughters or as sisters. In some cases, only one vixen becomes pregnant each breeding season, and when her kits are born, the other females in the group help to raise the litter. Fox groups can have complex social rankings, with dominant, aggressive animals taking precedence over more submissive individuals. For example, dominant group members, usually the male fox and breeding females, claim the choicest hunting areas. Each fox group has its own home range, which is the area where the foxes live and hunt. For foxes, home ranges are almost always territories that are defended against intruders, and resident foxes often use a complex system of scent-marking to demarcate the edges of their territories.

The size of a fox's home range varies with the species and also with the amount of food that is available. Each autumn, fox kits from that year's litters leave their families' home ranges to search for territories of their own; in winter, food is harder to find and foxes need a larger area in which to hunt, so home ranges expand. During these seasons, the screams and growls of foxes drift through the woods as neighbouring foxes encounter one another with increasing frequency.

When foxes meet at close quarters, they use their voices and body language to communicate. Foxes make a variety of noises, from a kind of chuckling "gekker" described by David Macdonald to a shrill scream. Each noise seems to have its own meaning. Subtle changes in the position of a fox's ears, tail and mouth also convey a wealth of information. Although much of a fox's communication is indecipherable to humans, we can understand some behaviours. For example, watch any group of fox kits for very long and you'll see the same "let's play" posture shown by your family dog—head and front of body held low, rear end up and tail wagging. If danger threatens, the same group of kits will be sent scrambling for the safety of their den by a sharp alarm bark from one of their parents.

Danger for foxes can come in a variety of forms. Many predators, including owls, large hawks, lynx, wolves, bears and other foxes, pose a threat. Garrott and Eberhart noted that several families of Alaskan arctic foxes abandoned their dens, apparently because of harassment by golden eagles. In Africa, jackals, hyenas and leopards prey on foxes, and in North

FACING PAGE: *Large prey, like this rabbit, may be partially or completely skinned before being eaten. The fox will eat the choice portions—heart, lungs, liver and forequarters—first, storing the stringy hindquarters for later.*
ART WOLFE

ABOVE: *A red fox stands poised at the entrance to its den, checking for any signs of danger. Fox dens are often easy to spot but should not be approached too closely, especially in spring, when human presence could make a vixen feel threatened enough to move her kits.* GLEN & REBECCA GRAMBO

RIGHT: *Making long- distance contact, a culpeo calls across a mountain plateau in South America. Foxes use a variety of vocalizations, from a soft whimper to a shrill scream, to communicate with each other.* WAYNE LYNCH

America, wolverines sometimes kill fox kits. Humans and their pets are the biggest problem for foxes, however. Domestic dogs often harass foxes and are one of the prime causes of death for fennecs and Bengal foxes. Humans exact a heavy toll through sport hunting, pest control, fur trapping and habitat destruction. Most foxes can expect to die, directly or indirectly, as a result of human activity. Life expectancy for foxes is extremely variable, depending on species and living conditions. In captivity, most foxes live a maximum of about thirteen years. In the wild, a fox surviving to its fifth birthday would indeed have something to celebrate.

Many of the world's foxes inhabit remote, sparsely inhabited regions, and much about them remains unknown. Wherever they live, they have one very important characteristic in common: they adapt. From dinner menu to den site, they are masters of improvisation, surviving by virtue of their flexibility in areas where extremely specialized animals could not. The foxes of North America and Europe are the most closely studied, partly because of their importance as furbearers and also because of their role in the spread of rabies (the second most common cause of fox deaths). Europe is home to the red fox and arctic fox. North America, with its larger landmass and greater diversity of landscapes, has the red fox, arctic fox, swift fox, kit fox and gray fox. These five species occupy a variety of habitats and provide a representative cross-section of what it means to be "foxy."

ARCTIC FOX

When biting winds scour the tundra, the arctic fox (*Alopex lagopus*) depends on its coat for survival. The long, fine guard hairs and lush underfur insulate better than any other mammalian fur, including that of wolves and polar bears. The arctic fox's ears are short and rounded like a bear's to minimize heat loss. Its eyes are heavily pigmented, acting as snow goggles against harsh glare from the snow and ice. Add short extremities and furry feet, and you have an animal well adapted to life in a frozen world.

This small fox (about the size of a large, fluffy Chihuahua) trots over the snow, covering tremendous distances. No other land mammals, except humans, travel farther. Individual foxes are known to have travelled over 1500 km (950 miles), and David Chesemore reports that mass migrations of arctic foxes have been observed in northern Europe and the former Soviet Union. During his 1893–96 trip across the polar pack, Arctic explorer Fridtjof Nansen saw fox tracks far out on the ice, 240 km (150 miles) north of Franz Josef Land, and tracks have been seen as far north as 88° north latitude and more than 450 km (280 miles) from the nearest ice-free land off Greenland.

The driving factor behind arctic fox movements is food. When shortages occur close to home, such as the periodic crash of the lemming population, foxes journey until they reach another food source. Even for the less-than-choosy fox, the search for food is a continual struggle.

During the winter, an arctic fox out on the sea ice may dog the steps of a polar bear, hoping for a scrap of seal. A frozen walrus carcass provides a communal feast for hungry foxes, who, according to David Chesemore, may claw through half a metre (1.5 feet) of ice before getting their teeth into the tough meat. Inland foxes rely heavily on lemmings and other rodents, adding ptarmigan or arctic hare when they can. These foxes often follow wolves, scavenging caribou carcasses, and will attack weak caribou calves themselves. Some arctic foxes stockpile supplies for the winter, caching food during summer's abundance. One such cache, mentioned by Ewer in *The Carnivores*, held thirty-six little auks, four snow buntings, two young guillemots and a large number of little auk eggs—probably enough to feed the fox for a month!

Throughout the long northern winter, the ringed seal provides essential food for the polar bear and scavenging arctic fox. Spring and summer give arctic foxes many more menu choices. In April and May, the seal pups are born and arctic foxes no longer have to wait for a polar bear to make the initial kill. Researcher T. G. Smith speculates that a keen sense of hearing and smell enables the fox to locate ringed seal dens buried under as much as 1.5 m

(5 feet) of snow, probably from as far away as 2 km (more than a mile) downwind. With deadly accuracy, the fox breaks through the roof of the den and seizes the pup within.

Colonies of nesting birds along the shore provide eggs, nestlings and the occasional unwary adult. If chasing birds is not to the fox's liking, it can stroll along the seashore and dine on the catch of the day. The arctic fox's sharp teeth crunch through crab shells and crack mollusks like walnuts. With catlike quickness, it snatches small fish from a tidal pool. Sea urchin and seaweed may complete the meal. Inland, foxes stalk the grassy areas where small herbivores feed. A quick pounce captures a lemming, hare or vole. If the fox lives near humans, it may visit their garbage dump. Here, the fox daintily licks the last trace of yogurt from a carton or picks bits of meat from yesterday's chicken dinner. If a fox pair is raising a large litter, they need all the food they can find.

When food is plentiful, the reproductive output of the arctic fox rises and litters of ten or more kits are not uncommon. When food is scarce, the vixen gives birth to a smaller litter, typically three to six kits. If the food shortage continues, the strongest kits will cannibalize the weaker ones. When food is very scarce, even this harsh strategy will fail as the adults abandon the kits in an attempt to ensure their own survival. If the parents survive, they will be able to produce more young next year.

Arctic fox dens usually appear as 1- to 4-m-high (3- to 12-foot-high) mounds on the open tundra. The foxes may also den in rocky areas, among dunes or in pingos (low ice hills pushed up in the permafrost), or they may take up tenancy in vacant Siberian marmot burrows. Arctic fox dens can be simple burrows with a few entrances or vast tunnel complexes with as many as a hundred entrances. The largest, oldest dens may have been in use for centuries. At the den entrances, the foxes constantly add organic material to the soil and aerate it with their digging. The resulting lush vegetation stands out on the otherwise dry tundra, making the dens easy to spot in the late summer.

The arctic fox itself is well camouflaged, but not always in white. The two colour phases, white and blue, can both be found in the same litter. The white form is white in the winter and brown in the summer. The less common blue form is usually a dark brownish- or bluish-grey in the summer and slightly lighter in the winter. The coat colour seems linked to the fox's habitat, with more blue foxes being seen in coastal areas, where less snow accumulates. In either colour, the winter pelt is valuable and the arctic fox is heavily trapped for its fur.

Arctic foxes are found throughout the Arctic or tundra regions of the world, from Norway eastward through Siberia and North America to Greenland. The metabolism of the well-insulated arctic fox increases only when the temperature falls below −40°C (−40°F), meaning that it requires less food than a red fox to keep going in cold weather. This advantage allows the arctic fox to exploit a gap in the red fox's range. At the lower margins of the arctic fox's domain, where food is more abundant, the red fox takes over.

FACING PAGE: *This arctic fox is well equipped to handle the cold. Its compact body, lush fur and small ears all help to keep heat loss to a minimum. The metabolism of an arctic fox doesn't begin to increase until the mercury drops below the −40°C (−40°F) mark, and this, combined with its smaller size, allows the arctic fox to survive on less food than a red fox would require in the same circumstances. This advantage lets the arctic fox exploit a gap in the red fox's range.* ALAN & SANDY CAREY

Dwarfed by the polar bear they are trailing, arctic foxes hope for a seal scrap if the bear makes a kill. During studies on polar bears carried out as far as 160 km (100 miles) offshore, arctic foxes or their tracks were seen wherever polar bears killed seals. All foxes are opportunistic feeders to some extent and will scavenge not only for carrion but for items as diverse as lambing afterbirth and food scraps in a human garbage dump. FRED BRUEMMER/VALAN

ABOVE: *Excavating a temporary larder, this arctic fox stores or "caches" an extra bit of food. Foxes around the world display this behaviour. By caching food when they have a surplus, foxes provide themselves with an insurance policy against lean times.* WAYNE LYNCH

RIGHT: *Playful jousting or serious aggression—only the foxes know for certain. The positions of their tails, ears and heads, as well as the sounds they make, are all part of their communication with each other. Foxes can also exchange information by vocalization or through scent-marks, which are usually made with urine or feces.* FRED BRUEMMER/VALAN

SWIFT FOX AND KIT FOX

To the south, breaks in the red fox's North American range are filled by two other specialized foxes—the swift fox and the kit fox. These small foxes are so similar that some authorities consider them to be a single species.

The swift fox (*Vulpes velox*) is a grasslands hunter and once lived throughout the west-central North American plains from the Texas panhandle up through the prairies of southern Canada. By the mid-1900s, human activity had drastically reduced the numbers of swift fox, especially in the northern part of their range, leaving only a small central population. After the 1950s, the swift fox began to reoccupy much of its historic range, and in 1983, Canada began a re-introduction program in Alberta and Saskatchewan. By the end of 1992, a total of 656 swift foxes had been released. There now appears to be hope for the re-emergence of a self-sustaining swift fox population on Canada's southern prairies.

When evening comes, the swift fox leaves its burrow and moves off across the prairie. As the light fades, the fox begins to hunt for the mice and gophers that are its main food. It will tackle prey as large as black-tailed jackrabbits—quite a challenge for a fox not much bigger than a house cat! With a good jump, it is possible for the swift fox to overtake a fleeing jackrabbit, running at speeds of up to 60 km/h (about 40 miles per hour). Flowing like a tawny liquid through the prairie grasses, a racing swift fox closes in on its prey, its feet hardly seeming to touch the ground. Its speed and ability to turn on a dime make it a lethal predator of small mammals.

Although rabbits, mice and ground squirrels are the staples in the swift fox's larder, like any efficient forager, it takes advantage of seasonal foods. Beetles, grasshoppers and other insects can make up a large portion of the fox's diet when they are plentiful. Winter-killed deer and other carrion may be important food sources, particularly when other food is scarce. Birds, lizards, grasses and fruits round out the swift fox's diet.

As the sun rises, the fox returns to its burrow to spend the day sleeping. Adult swift foxes live in pairs and may mate for life. Their yearly courtship takes place sometime from late December to early February and pups are born from March through May. The average litter consists of four or five fuzzy, golden babies. Swift foxes dig their own burrows or take over the holes of badgers, prairie dogs and other animals. Pups and adult foxes may be seen outside their burrows sunning and resting but don't venture far during the day. Unlike most foxes, who only use their dens while they are raising kits, swift foxes use their burrows year-round. A swift fox family may occupy as many as thirteen dens in one year, moving because prey is scarce in the den area or because of a buildup of skin parasites inside the den.

FACING PAGE: *Once again at home on the range, this swift fox family is part of the Canadian re-introduction program, which is putting foxes back into areas from which they had been exterminated.*
WAYNE LYNCH

The little swift fox lives only three to six years in the wild and has many enemies. Great horned owls, bobcats and golden eagles take swift fox pups. Renardo Barden, in *Endangered Wildlife of the World*, writes that wolves were tolerant of the small foxes, but their elimination by humans opened the way for more coyotes and red foxes. Closer in size to the swift fox, these two predators often compete with the swift fox for prey. Coyotes also pose a more immediate threat, and the swift fox must often flee at top speed to the nearest bolt hole to avoid being killed. As for most foxes, however, the worst problem for swift foxes has been humans, who have trapped them and destroyed their habitat.

Superbly adapted to desert life, the kit fox (*Vulpes macrotis*) lives in the arid and semi-arid zones of the western United States, ranging southward into northern Mexico, including the Baja Peninsula. Like the arctic fox, the kit fox has hairy feet, heavily pigmented eyes and a pale, thick fur coat, but the kit fox puts these features to different use. The hairy feet enable it to travel over hot sand, the extra pigment protects its eyes from the fierce desert sun, and the fur coat insulates it against both the heat of the day and the chill desert night.

Like many mammals that dwell in deserts, the kit fox is nocturnal, avoiding the heat of the day, and it has huge ears, which act as radiators to get rid of excess body heat. As it hunts during the cooler hours, the kit fox's remarkable ears help it to find tasty insects by functioning like parabolic microphones to gather and focus tiny noises. Besides insects, the kit fox seems partial to kangaroo rats, but it also eats mice, voles, birds and cactus fruits. Although a kit fox rarely weighs more than 1.8 kg (4 pounds) and is slightly smaller and slower than the swift fox, it too hunts black-tailed jackrabbits. This varied diet gives the kit fox all the moisture it requires, freeing it from the need to find a source of drinking water—an important adaptation to its desert home.

Like swift foxes, kit foxes use their underground dens throughout the year, perhaps partly to avoid coyotes. Choice denning areas may look like fox housing developments, having groups of dens, each with several entrances. Where natural dens have been destroyed, kit foxes improvise, using well casings, culverts and abandoned pipelines. Kit fox pairs share a strong bond and are believed to stay together throughout the year. For the kit fox, mating takes place sometime from December to February. About fifty days later, four or five tiny babies are born, each weighing only about 40 g (1.4 ounces). While the mother is nursing, she rarely leaves the den, depending on her mate to bring her food. Kit fox families, like swift foxes, may change their home address several times during the summer.

Kit fox pups first venture outside the den when they are about one month old. Two or three months later, they begin hunting with their parents. Like other species of fox, the young foxes leave to seek out new territories when autumn comes.

FACING PAGE: *Although the red fox has the largest geographical range of any carnivore and is, in some cases, expanding that range, not all foxes are doing so well. Threatened by habitat loss and continued encroachment by humans, the San Joaquin kit fox is classified as endangered and its future is uncertain.* ERWIN & PEGGY BAUER

GRAY FOX

The gray fox (*Urocyon cinereoargenteus*) is at home hunting mice in the Venezuelan scrublands or chasing cottontail rabbits through the forests of the southern United States. With a range reaching northwards to the Great Lakes and extending south through Central America into Venezuela and Colombia, gray foxes occupy a wide variety of habitats. They prefer wooded areas, frequently in broken or rocky terrain, but are found in citrus groves, chaparral, railroad rights-of-way and dry, open country. Brushy areas along the boundaries of clearings provide good small-mammal habitat and ideal hunting areas for foxes. Gray foxes take advantage of this edge effect and live near city margins and cultivated areas across their range. The closely related island gray fox (*Urocyon littoralis*) is found only on six of the Channel Islands off the southern California coastline and is listed by the state of California as "threatened."

Depending on where gray foxes live and what foods are locally available, their diet can vary tremendously. In the autumn, they may enjoy a largely vegetarian diet. Corn, apples, peanuts, persimmons, hickory nuts and grasses are just a few of their favourites. When plentiful, insects make up as much as 40 per cent of the gray fox's diet, and small mammals, especially rabbits and mice, are important throughout the year.

The gray fox has an unusual trait that sets it apart from other canids. If a red fox is pursued, it often runs until it finds a way to shake its pursuers. If a gray fox is chased, it prefers to slip down a hole or look for a tree to climb. Like a cat, the gray fox grasps the tree trunk with its front paws and pushes itself upward with scrambling back feet. Once up, it may leap from branch to branch. A close look at the gray fox's anatomy shows that its forearms have more rotational mobility than those of other Canidae, perhaps improving its ability to climb. Descent from the tree is also catlike—head first on a sloping tree or back feet first on a vertical trunk.

Even when they are not being pursued, gray foxes climb trees to rest or forage there. North American writer and naturalist Ernest Thompson Seton wrote of seeing a gray fox resting quietly in an abandoned hawk's nest. Many people have watched gray foxes climb straight up trunks with no low branches, and gray fox kits can scale a vertical trunk without assistance by the time they are one month old.

Gray fox kits are born in dens placed in burrows, rocky crevices and a host of other places. Victor Cahalane reports that one enterprising vixen raised her litter in a discarded 40-L (10-gallon) milk can! The climbing ability of gray foxes allows them to place dens in hollow trees almost 8 m (25 feet) above the ground. The average litter size is four, but gray

FACING PAGE: *The gray fox is possibly the only member of the dog family that would feel comfortable in this situation, explaining why it is sometimes called a tree fox. Ranging from near the United States/Canada border south through Mexico and Central America into Venezuela and Colombia, the gray fox hunts in a wide variety of habitats.*
LEONARD LEE RUE III

foxes may have only one or as many as ten kits. The babies are born from mid-March to mid-June, arriving earliest in the southern portions of their range. At this point, males in most fox species play a large role in caring for the nursing vixen and raising the kits, but gray fox fathers are reportedly less attentive. When the pups are three months old, they start travelling away from the den with their parents. A month later, the kits are hunting on their own. Gray fox kits remain in their parents' home range longer than kits of other foxes, staying until January or February of their first year.

FACING PAGE: *Although the rust-coloured portions of its coat may lead some people to confuse it with a red fox, the predominant salt-and-pepper coloration and the black dorsal stripe on its tail make it clear that this is a gray fox.*
DANIEL J. COX

RED FOX

For most of us, the word *fox* conjures up the image of a playful red animal with an intelligent expression on its pointed face. The red fox (*Vulpes vulpes*) is the world's most familiar fox, which isn't surprising, considering its wide range and frequent interaction with people. Today red foxes occupy a greater geographic range than any other carnivore, living on five of the six continents, from the fringe of the northern tundra to the edge of the Sahara.

The key reason for the wide range of the red fox is its amazing adaptability. All foxes share this trait to some extent, but none can match the flexibility of the resourceful red fox. For example, unlike many other animals, red foxes are able to cope with and take advantage of human changes to the environment. After World War II, red foxes began making their homes in the bombed-out ruins and abandoned buildings found in many European cities. Some of the densest red fox populations in North America are in farm and dairy communities, and red foxes are frequently seen trotting through the streets of urban areas, seemingly undisturbed by the hectic pace of life around them.

Another form of adaptability is revealed when you compare the size of red foxes living in various locations and habitats—northern red foxes are consistently heavier than their southern counterparts. Bigger animals have less surface area in proportion to their body mass than small animals do; therefore, they give up precious body heat less easily. Bigger animals can also store more energy as fat, which helps them stave off winter's lean times and also acts as extra insulation.

Although we usually think of red foxes as bright red, an adult fox can be any shade from pale yellowish-red to a shimmering silver-black. For example, silver foxes, whose pelts are popular with furriers, are red foxes with silver-tipped guard hairs throughout their mostly black coats. A cross fox is not a cross-breed of any kind but is simply a red fox with a dark line of fur down its back and a second dark line across its shoulders. All three colour phases—red, cross and silver—may be found in a single litter. Captive foxes have been selectively bred for other colours, but wild foxes usually appear in only two other variations: a Samson fox is a genetic mutation with no guard hairs, only soft, tightly curled underfur, and a bastard fox is a rather smoky colour, halfway between red and black.

Whatever their colour and wherever they live, red foxes centre their lives on the daily hunt for food. If they kept pantries like ours, instead of "Flour," "Sugar" and "Rice," containers holding red fox staples would be labelled "Voles," "Rabbits" and "Mice." These basics make up a major portion of their diet, but if you were to set yourself the task of writing down every food red foxes are known to eat, you would be compiling a very long menu

FACING PAGE: *Leonard Lee Rue III writes that if you are lucky enough to view a red fox standing in the sun against a pristine snowscape, you are viewing the* beau ideal *of the animal world. I agree.* ALAN & SANDY CAREY

indeed. Each season and locale offers different choices for the catholic tastes of the fox.

Fruit such as strawberries, juniper berries, raspberries, rose hips, plums and grapes may be the main food during the seasons when they are abundant. Eggs, birds, winter-killed fish, deer carcasses, bumblebees and grasshoppers are also on the list, which is amazing in its variety. Anglers out gathering earthworms after a night rain may have a competitor they never considered. Red foxes hunt for the worms, listening for the rasp of their bristles on the wet grass.

Besides the standard mousing leap and rabbit stalk, red foxes have two very unusual hunting techniques. Desmond Morris, in *Animalwatching*, writes that for hundreds of years, tales have been told of a clever fox who played dead in order to catch the birds that came to feed on its carcass. Pictures in medieval bestiaries showed the fox lying upside down with its eyes closed and its tongue hanging out as the birds gather round. After waiting until the last possible moment, the fox jumps up, catches the birds and eats them. No one really believed this could happen until, in 1961, a Russian movie maker captured a fox's thespian activities on film. At first the film shows the fox just lying in the grass, limp and unmoving, eyes closed. A carrion crow slowly approaches and prepares to enjoy a feast. Suddenly, the fox springs round, catches and kills the crow, then settles down to eat. Ancient legends may contain more truth than we think!

The red fox displays the same kind of intelligence when it tolls for ducks. Noted North American naturalist and wildlife photographer Leonard Lee Rue III describes how a fox will pick up a stick and play with it along the shore, in full view of the ducks. The curious ducks swim closer to see what is happening while the fox studiously ignores them. After a bit, the fox seems to tire of its game and disappears into some brush or reeds. Any ducks foolish enough to come up onto the shore to investigate soon find themselves making a meal for the fox. So common was this method, and so successful, Rue says, that hunters along the New England coast bred special, medium-sized yellowish dogs and trained them to toll the ducks in close to shore, giving the hunters an easier shot.

Besides their daily routine of looking for food, red foxes follow a yearly cycle that periodically adds a new urgency to their hunting. Red foxes mate in the middle of winter, following some inner calendar that appears to be tied closely to day length. The actual calendar date for breeding changes with location. For example, foxes in the southern United States mate and have their kits earlier than northern foxes. And in Australia, of course, everything is switched around by six months.

About two months after the foxes mate, a litter of between one and thirteen pups is born. Both parents care for the pups, and sometimes they have a "helper," usually a vixen from a previous litter. The helper vixen may be encouraging the continuance of her own genetic line by caring for closely related kits, or she may just be sharing in the abundant food that the parents bring to the den. In at least one case, cited by J. W. Sheldon in *Wild Dogs: The Natural History of the Nondomestic Canidae*, a nonbreeding female raised a litter of kits whose mother had died. David Macdonald indicates that in large red fox social groups, usually only the dominant vixen becomes pregnant and raises pups. The other vixens in the group are generally friendly towards the kits and spend time guarding and sleeping with them.

Research has shown that socializing between red foxes is more common and complex than was previously thought. Although they hunt alone, red foxes often spend at least part of their evening exchanging messages with each other. When they are close enough, they use a variety of sounds to communicate. Whining, shrieking and a variety of low, throaty noises can be heard when one fox encounters another. Long exchanges between fox pairs float through the night air, particularly during their winter courtship.

Red foxes have a complex system of scent-marking that functions as a notice-board for other foxes. Both males and females out on nightly patrol frequently stop to mark with urine or feces such features of the landscape as a corner fence post, stump or rock or even a particular clump of grass. If it is visiting someone else's territory, a red fox urinates much more frequently, either because it wants to advertise its presence or because it is uneasy about being on someone else's turf. Neighbouring foxes appear able to recognize each other's scent marks, and by posting messages along territory boundaries that clearly indicate ownership, vicious battles can be avoided. This allows foxes in adjacent groups to concentrate their time and energy on obtaining food and raising families.

Red foxes have an assortment of scent glands that leave additional aromatic messages. The real intricacies of the red fox's language of smell are still a mystery, although some of its marking behaviour is understood and the source of its scents is known. A glandular area of skin under a patch of dark hair on the upper side of the tail emits a "flowerlike" fragrance, giving rise to the common name of violet gland. It is particularly noticeable on the woolly grey tails of baby foxes. Anal sacs hold a bacteria-rich liquid that may be sprayed onto the fox's droppings. Researchers know that males add this extra scent more often than vixens, but they don't know why, nor do they know why it is only done sometimes. Red foxes also have scent glands in the skin near the chin and angle of the jaw, and between the toes and pads of their feet. The pleasant odour that comes from these paw glands is the scent that hounds follow when pursuing a fox.

Besides fox hounds, red foxes have several natural enemies to watch out for, including cougars, lynx and bears. They are especially cautious when coyotes are around, since coyotes seem to have a particular antipathy towards foxes. A. B. Sargeant and S. H. Allan report that one researcher observed two coyotes travelling together through a hay meadow. The coyotes startled a red fox into flight, then chased and quickly killed the fox. Later that day, the

researcher found another freshly killed fox whose wounds indicated it had also been attacked by coyotes. The fact that the foxes were not eaten indicates that something other than the typical predator-prey relationship may be occurring. Perhaps the coyote views the red fox as a competitor to be eliminated.

For red foxes, as for all foxes, fortunes change as often as the wind and weather. They must create their own luck through their willingness to use what's available, their expertise as hunters and their habit of storing food against lean times. Although their individual triumphs and failures, for the most part, go unrecorded, their success as a group is clear—tonight there will be foxes gliding through the shadows.

FACING PAGE: *Holding a freshly captured vole in its mouth, this red fox stops to urinate. Researchers observing this type of behaviour speculate that the scent-mark tells other foxes that searching for voles here is pointless.* DANIEL J. COX

PAGES 34–35: *A small hunter alone on a vast plain of ice and snow, the arctic fox has adapted successfully to one of the most demanding environments on Earth. Adaptability to changing seasons and food supplies has helped foxes find homes on all but one of the six continents in habitats as different as tropical rain forest and coastal desert.* ART WOLFE

Chapter 2 SEASONS OF THE RED FOX

Just a two-hour drive north of where I live lie the woods and lakes of Prince Albert National Park. Here, generations of red foxes have grown up free from the threat of trapper and hunter, making it an ideal setting to learn what it is like to be a red fox. The following story of a red fox family, which is based on my observations and those of others, offers a glimpse of behaviours that are typical of red foxes around the world.

SPRING: A TIME OF NEW LIFE

Outside, the March wind is howling and snow still covers the ground, but inside the den, close to the mother fox, it is warm. The vixen cleaned out several dens before finally choosing this one. It met her requirements: several entrances, dirt that is easy to dig but stable and well drained. There is cover nearby and water close at hand.

Tired from her recent exertions, she rests as her five new kits suckle contentedly. The kits are small, fuzzy, grey balls, weighing about 110 g (4 ounces), about the size of a mole. Their tiny tails already display the characteristic white tip. The kits are blind and helpless, relying on their mother for food and warmth. For now she remains with them almost constantly, while her mate brings food to the den.

About two weeks after they are born, the kits' eyes open and they begin to take an active interest in their world. The vixen begins hunting again, searching the grass and weeds near the den for mice and voles. She returns to the den at intervals to play with her kits and to

FACING PAGE: *These young red fox kits are sleeping soundly, reassured by the warmth and scent of their mother. David Macdonald suggests that without the thermal blanket provided by the vixen, newborn fox kits would have a tough time surviving. This is especially true in the northern reaches of the red fox range, where kits come into a world still blanketed in snow.* DANIEL J. COX

ABOVE: *Only two days old and still blind, this red fox kit displays the white tail tip it will carry throughout its life.* LEONARD LEE RUE III

nurse and clean them. Her milk is about three times richer than cow's milk, and the kits grow quickly.

The female spends more time near the den than does her mate. Today, as always when he returns from hunting, she greets him delightedly, crouching low and waving her tail over her back. In his haste to return her greeting, he drops his cargo of voles and has to retrieve them before moving on to the den. Although the kits will not be ready to eat solid food until they are about a month old, they grip the meat eagerly with their tiny teeth and suck on the juice, developing a taste for fresh vole or hare.

A couple of weeks later, a small grey snout cautiously pokes out of the entrance to the den. Conditions satisfy the fox equivalent of "All Clear," and out tumble the kits on wobbly little legs, chasing each other and their own fuzzy brush tails. They are a month old now, and their round blue eyes look at the world with insatiable curiosity. Every new sight or scent has to be thoroughly explored. Today they are timid, appearing only because their mother is near, but it won't be long before they start to come out on their own.

The vixen is looking very ragged. Her long winter coat is bleaching in the sun and coming out in bunches, giving her a moth-eaten appearance. Nursing, too, has taken its toll. The kits' sharp teeth have left sores around her nipples and torn clumps of hair from her stomach. It is time for her to wean the kits, and she lies down when they try to nurse. The kits, thoroughly frustrated, turn hopefully to the male. They fruitlessly grasp tufts of hair, and he stands for just so much poking and prodding before trotting off.

The kits are now eating solid food, and the family has established a routine when food is brought to the den. A short distance away, the parent calls softly to the kits, making a rather mumbled sound through a mouth full of mice. The kits rush out and greet the adult by wriggling low on their stomachs and wagging their tails. One of the kits then licks and bites at the corners of the adult's mouth, asking for food. As a rule, the first kit to approach is the first one to be fed, ensuring that when food is scarce, the strongest of the litter gets the most.

Getting food doesn't necessarily mean that the kit will keep it. Hungry sisters and brothers wrestle and bite, sometimes creating a tug of war in which a vole is torn in two. Kits take what food they can and either run into the den to escape the others or stand and fight to defend their share. More aggressive kits steal food from their less dominant littermates with no interference from their parents. This seemingly unfair behaviour has evolved to help preserve the species. Stronger foxes have a better chance of surviving and reproducing.

The kits are spending more time outside the den now, and their coats are changing from charcoal grey to light tan. Their fur is still very soft, since they don't yet have the long guard hairs of adult foxes. No longer blue-eyed babies but golden-eyed juveniles, the kits are larger and more coordinated. The days are growing longer. It is the end of May, and the kits are about to experience their first summer.

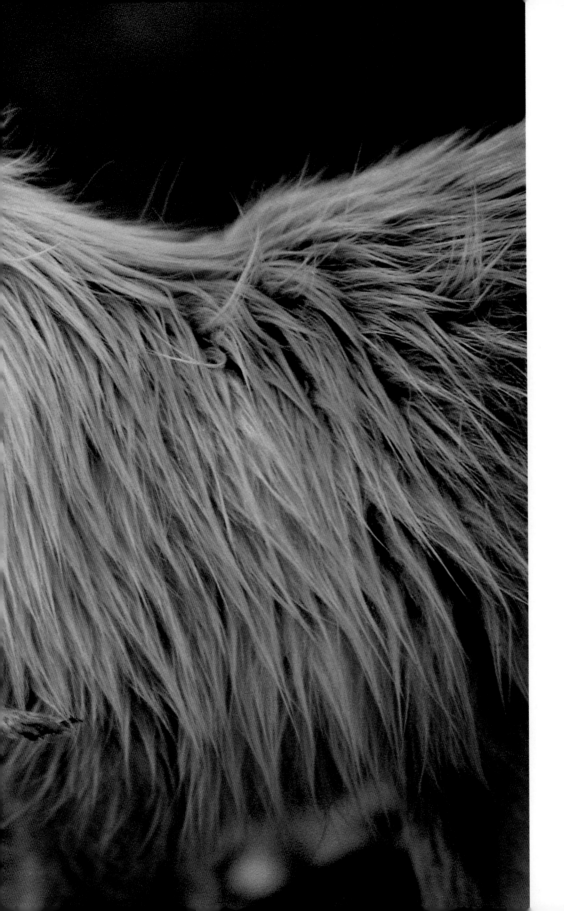

A mouthful of arctic ground squirrels will be met with appreciation from the hungry family waiting for this red fox. As the kits get older, their parents will begin leaving food farther and farther from the den, encouraging the kits to forage. LEN RUE JR.

SEASONS OF THE RED FOX 41

LEFT: *Still in their grey baby fur and probably on one of their first forays from the den, these red fox kits are interested in everything around them.* ALAN & SANDY CAREY

ABOVE: *"Mmmm—good!" This six-week-old red fox kit seems to find gray squirrel to its liking. Although weaning begins at about five weeks, kits often begin chewing and sucking on food before then.* LEONARD LEE RUE III

ABOVE: *A young "cross" fox,*
actually just a colour phase of the
red fox, licks at the corners of its
mother's mouth, begging for food.
ALAN & SANDY CAREY

RIGHT: *"The more we come in*
contact with animals, and observe
their behaviour, the more we love
them, for we see how great is their
care for their young."—*philo-*
sopher Immanuel Kant. TOM &
PAT LEESON

Curiosity overcoming caution, this red fox kit risks a closer look at the interesting creatures (photographers) near its den. Adult foxes show this fascination with new objects too. The pampas fox of South America sometimes collects and stores objects like strips of cloth and leather. Unfortunately—for they are intensely hunted—pampas foxes are also curious about humans and may remain completely motionless when people approach. ALAN & SANDY CAREY

LEFT: *It's only a mock battle, but already these red fox kits are establishing a pecking order among themselves. The same behaviours can be observed when two adult foxes encounter one another.*

DANIEL J. COX

ABOVE: *With its attention focussed on a bug in the grass, this kit is about to get a rude awakening! Sneaking up on each other seems to be a favourite game for kits.* DANIEL J. COX

SUMMER: A TIME OF ABUNDANCE

The kit focusses her attention on the grass ahead of her. Suddenly she pounces, stabbing forward with her forepaws. Carefully, she lifts one paw to peer underneath. Out springs a grasshopper, and the kit patiently begins the game again, stalking her prey with care. Another pounce, and this time the game is over, ended by a quick snap of teeth.

Learning to hunt is the most important lesson of the summer. The kits hone their stalking skills on insects, while the vixen and her mate continue to supply food for their growing family. Along with the staple voles, mice and hares, the foxes enjoy berries, an occasional songbird, the leftovers of a camper's picnic and scraps from a road-killed deer. The parents are spending more time away from the den and no longer bring food right to the den entrance. Instead they drop it some distance away, encouraging the kits to forage. Meanwhile, the kits' pounces grow more confident and coordinated, and they regularly snap up crunchy beetles and grasshoppers, making insects a large part of their diet.

Although they do not know it, this first summer will be the most carefree time the kits will ever experience. Their stomachs are full, the warm days and cool evenings are meant for play, and the kits seem to be in constant motion. With small growls, they tumble over and over, stopping with an abrupt yelp when someone's sharp little teeth bite too hard. When one kit tires and lies down to rest, another jumps into action.

If no one else wants to play, almost anything makes a good toy. A stick is subdued by jumping on it and then shaking it fiercely. Once dispatched, the "dead" stick is carried around by the victorious kit and displayed proudly to his siblings. Nearby, a butterfly hovers just out of reach of another kit, who watches it with almost cross-eyed concentration. Tired of amusing himself, the kit with stick drops it and sneaks up on a sleeping sister. He launches a surprise attack, and the result is a pell-mell chase through the grass that eventually turns into a game of tag. These games are played for fun now, but the skills the kits are learning mean the difference between life and death for them during the coming fall and winter.

As the kits get older, the vixen or male takes them on short hunting expeditions. They follow their parents and learn to forage for many different kinds of food. Sometimes when the adult catches a vole, it will release it again, letting the kits try out their hunting skills. At first, the parent leads the kits back to their den, but soon they return on their own. As the kits begin hunting more independently, their trips away from the den grow progressively longer.

For the kits, their play-filled summer is nearing its end. Now, in early August, the red guard hairs that began growing into their buff coats in June have turned them into small replicas of their flame-coloured parents. Soon their summer of learning will be put to the test as the kits leave their familiar territory and seek out homes of their own.

FACING PAGE: *Full of energy, a two-month-old red fox kit romps in the sun. Although the summer is a time for play, the skills kits practise in their games will be of vital importance for surviving on their own.* THOMAS KITCHIN

ABOVE: *An unwary raven becomes a meal for a red fox family. Birds and their eggs are rarely a major diet item for foxes, except when concentrations of ground-nesting birds provide a rather easily accessible food source.*
ERWIN & PEGGY BAUER

RIGHT: *Summer evenings are just right for stalking bugs that seem to hang tantalizingly out of reach. When young foxes are learning to hunt, insects can make up a large portion of their diet.*
GLEN & REBECCA GRAMBO

AUTUMN: A TIME OF CHANGE

The male fox trots along the edge of the gravel road, stopping often to listen to noises in the tall grass beside him. He stands motionless, listening intently until he pinpoints the source of the noise. Suddenly, he launches himself with a catlike spring, landing and jabbing into the tall grass with his forepaws to trap the prey he cannot see. He had missed before, but this time his head lifts triumphantly, showing the mouse in his mouth.

It has been a good morning for the fox. He has eaten his fill of berries and voles, so this mouse can be cached for the future. Still holding the mouse in his mouth, the fox carefully chooses a site and excavates a mouse-sized hole with his front paws. He doesn't spray the dirt around as dogs do when they dig, but instead carefully piles the dirt beside the hole. The fox drops the mouse into the hole and then, bit by bit, pushes dirt over it, tamping down the layers with his nose. A last brush of whiskers and the hole disappears, completely camouflaged in the forest floor.

Larders such as this are scattered throughout the fox's territory. An adult fox requires about half a kilogram (a pound) of meat, or its equivalent in other food, each day, and when prey is scarce, he may rely almost exclusively on his caches. In these woods, magpies, bears and martens sometimes rob his stores, but other foxes are perhaps the greatest threat. A fox must take great pains to hide his extra provisions if he wants them to be there when he needs them.

Summer is over, and the fox family is breaking up. The young males leave first, driven by the urge to establish a territory of their own and to avoid increasing conflict with their father. The young females stay on for a while, but soon they too leave the home range. The vixen remains in the area, and the male notes her scent mark on an old deer carcass.

One of the young males travels farther than the others. The first area he moves into is already taken by another male fox. The young fox notices the scent marks that serve as an "Occupied" sign and is nervous. He moves cautiously and stops to urinate frequently. Suddenly he comes face to face with the rightful owner of the territory. The older fox lunges at the young male, making a low, rasping noise deep in his throat. The two foxes roll about in a furious mass, bones slamming, jaws snapping. With a sharp yelp, the young fox hurls himself to his feet and races away, pursued by the older male. The chase continues until the older fox stops, bristling with apparent anger, while the young male flees through the woods. He won't venture this way again.

So far, food is not a problem for the young foxes. The cool nights chill insects into torpor, making them easy prey. There are still plenty of berries, and small rodents are plentiful.

FACING PAGE: *Like many growing youngsters, fox kits collapse from a whirl of activity into sudden naps. After a quick recharge, this kit will be ready for more adventures.* ULF RISBERG/ NATURFOTOGRAFERNA

Fox families break up in the autumn as the young foxes leave their home ranges and search for territories of their own. The distance the young foxes travel varies with habitat and species, but some exceptionally long treks have been reported—one red fox journeyed 500 km (300 miles). Trappers and hunters concentrate their efforts at this time when foxes are on unfamiliar terrain and thus are more vulnerable.

ALAN & SANDY CAREY

The young male spots a snowshoe hare and begins a slow, deliberate stalk. His belly low and whiskers quivering with concentration, he carefully lifts and places each foot with silent precision. Without warning, the fox's cover is blown as the hare catches sight of him. Both animals explode into flight and tear into the brush at top speed. The chase ends abruptly with a shrill scream as the fox catches and kills the hare. After eating his fill, the young male carefully makes several small caches, storing the leftovers for later.

Very early one morning, one of the young vixens is hunting near the lake. Her nose tells her that people are nearby. People often mean food, so she moves closer to investigate. When she reaches the campsite, she stands very still and lifts her nose to have a good sniff—definitely food! She pads silently across to the picnic table. Underneath are a few potato chip crumbs, but no other food is in sight. She places her front paws on the bench and stretches up for a better look—nothing. She sidles closer to the tent and glimpses something white peeking out from under the fly. With one quick motion she pulls it out and looks at it.

The object isn't food, but it is certainly interesting. It is shaped like a fat cylinder with a hole through the middle and the outside comes off in layers when she rolls it on the ground. Grabbing some of the stuff in her mouth, the vixen trots off to one side. The toilet paper trails behind in a white ribbon. As the paper unwinds, the vixen begins to play in earnest, stopping only when a noise comes from the tent. She quickly leaves the campsite, thereby missing out on the expressions of the campers when they look out of their tent to see drifts of toilet paper covering the ground.

These are the last campers the vixen will encounter this year, for it is late autumn. The chilly air carries the bugling calls of elk, and golden aspens climb the hillsides. Already the first frosts have come and gone, and black Vs of geese thread across the skies. Snow will soon follow, and the young foxes will face the harsh test of winter.

FACING PAGE: *Fox pairs form close bonds and often play together during their courtship. Here a red fox gently mouths its mate. The courtship period allows this usually solitary animal to become accustomed to the presence of another.* ALAN & SANDY CAREY

WINTER: A TIME OF SOLITUDE

The young vixen has hunted through the night and into the first hour after dawn. Now her stomach is full and she is tired, ready to curl up for a rest. She buries her nose in the long fur of her tail and sleeps, kept warm by her lush winter coat. When she awakens, fluffy snowflakes are tumbling from the grey sky and beginning to form white patches on the ground. The vixen lifts her face and feels the snowflakes on her whiskers. She snaps at the flakes but can't seem to catch them. She licks experimentally at one of the snow patches and is puzzled by its smell and taste. As she moves off through the silent woods, the vixen continues to look around at the growing blanket of her first snowfall.

With the snow comes an end to easy hunting. Many of the small burrowing mammals that the foxes have been eating are tucked away, hibernating through the winter. Snowshoe hares still hop through the woods and voles trundle through tunnels under the snow, but competition for the available food is fierce. Martens, fishers, weasels and owls all have their sights set on the same prey as the foxes.

The "listen and pounce" technique that proved so effective in the long summer grasses serves the foxes well in the snow. The male fox waits on a path, head cocked to one side, listening. He can hear the minute scratchings of a vole scuttling along its tunnel trackway. The fox springs, plunging his face and forepaws downward, sending up a spray of snow. His reward is a fat vole, which he tosses into the air and catches several times before finally crunching it between his jaws. The snow is not yet too deep, and the male fox doesn't have much difficulty in finding food.

Two of the kits have not been so lucky. One of the young males was struck by a car and killed while scavenging a road-killed deer. Another travelled outside the safe boundaries of the park and was driven by hunger to take the bait in a trap. Three of the five kits remain alive.

The two young females are both doing well. Their hunting skills keep them supplied with food. On a bitterly cold day, one of the vixens watches from a distance as a ruffed grouse plunges into a snowbank, seeking shelter. Insulated under the snow, the bird doesn't hear the quiet approach of the fox and is taken unaware by the vixen's sudden attack. After a short flurry of fur, feathers and snow, the action is over. The grouse makes a large meal, and the sated fox painstakingly caches the remains in the snow.

The remaining male kit finally found an unoccupied territory and now patrols its borders on his hunting trips. He stops frequently to deposit urine scent marks on rocks, trees, snow mounds and old kills, checking carefully each day for any indication of intruders.

In the old home range, the vixen and male fox have crossed paths once in a while but without any prolonged interest. Now that begins to change. Throughout the forest, a musky, "skunky" odour floats from fox scent marks. Soon, when the vixen meets the male, they travel together for a while. At first, they spend only short times together, but by the middle of January they are together almost constantly. The two foxes touch frequently and call to each other when apart.

Their courtship is consummated in the bleakest part of the northern winter, during the few days when the vixen is receptive to the male's advances. When they mate, they are bound together for about fifteen minutes by an enlargement of the male's penis that creates a "tie" similar to that seen in dogs. The reason for this is that the male takes several minutes to finish his ejaculation. The tie prevents the animals from separating until after the male has ejaculated, greatly increasing the chance that the vixen will conceive. After mating, the two foxes go their separate ways but remain within their home range, hunting and marking their territory.

The three remaining kits, who are kits no longer, have all found mates and territories of their own. The days slowly grow longer, and the ice on the lakes begins to groan and break. Throughout the northern forests, red foxes are choosing and cleaning out dens. And written across the snow in paired fox tracks is the promise of new life and another year of foxes.

For a red fox, life is short and intense. Only a minority of kits make it through their first year, and rare indeed is the fox that lives to see its fourth birthday, particularly in areas that are heavily hunted and trapped. Why did this litter have a relatively high survival rate? The park offered protection from the main agent responsible for this heavy mortality—humans.

RIGHT: *Sometimes fox pairs return to the rough-and-tumble games of their youth.* ALAN & SANDY CAREY

FACING PAGE: *This rabbit will make more than one meal for the fox, who will eat its fill and then cache the leftovers under the snow. Later, the fox will be able to relocate its larder by using a combination of scent and memory.* LEONARD LEE RUE III

62 FOXES

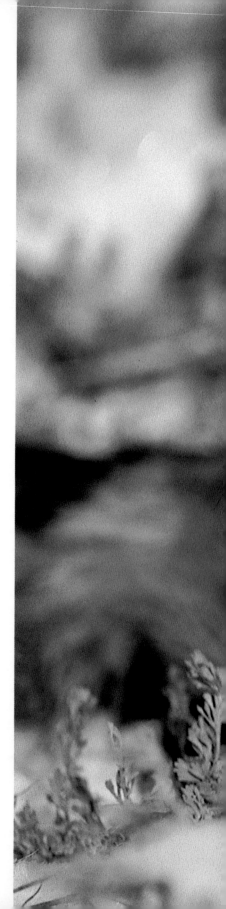

RIGHT: *Faithful mates and good parents, this red fox pair will work together to raise the kits born as a result of their winter courtship. Should one of them die, the survivor may attempt to provide for the family on its own.*
ALAN & SANDY CAREY

PAGES 66–67: *Foxes play a role in the folklore of people around the world. In Lappland, where arctic foxes like this one were believed to be the source of the northern lights, the earliest known illustration of the aurora shows a fox-hunting scene. This curious twist means that the hunter is able to see the fox because of illumination provided by the fox.* WAYNE LYNCH

Chapter 3   HUMANS AND FOXES

FOXES IN MYTH

The crowded marketplace was a riot of noise and movement. Taking advantage of the confusion, a bold crow swooped down, snatched a bit of meat from one of the stalls and flew to a nearby tree with his booty. A fox spied the crow settling onto his perch and resolved to get the meat for herself. Once beneath the crow, the vixen began praising him in buttery tones, admiring his size and great beauty. Such a magnificent bird deserved to be king of all the birds, purred the fox, and would be, if only the crow had any kind of voice. As the crow opened his mouth to demonstrate that he did indeed have a voice, the meat fell down to the waiting fox. "If only, dear crow, you had any sense at all," laughed the vixen, "you would truly be fit to be king!"

—A retelling of "The Fox and the Crow" by Aesop

Aesop told this tale more than 2500 years ago and the ending, with the fox's intelligence winning the day, is found in many tales of that time. The ancient Greeks were not alone in their interest in foxes, which play an important role in the folklore of virtually every civilization that has had contact with them. They have been celebrated as creatures of wonder and as beings possessing great intelligence and cunning. They have also been strongly associated with sexual attractiveness and, like many other animals active mainly at night, with evil and death.

Fabulous foxes are often linked with gods and the heavens. In Finland, the aurora borealis was given the name *revontulet*, which means "fox fires." Legends tell that the northern lights come from the glittering fur of a fox running across Lappland's mountains. Japanese

FACING PAGE: *This fox head finial, beautifully crafted of copper and silver by the Moche people of South America around* A.D. *500, probably fitted over the end of one pole of a dignitary's litter.*
MICHAEL CAVANAGH & KEVIN MONTAGUE/INDIANA UNIVERSITY ART MUSEUM

foxes are venerated as messengers of the benevolent Shinto rice goddess, Uka no Mitama. The creation myth of the Achumawi people of California speaks of a small cloud appearing out of nothing and condensing to form Silver Fox. When Coyote appears from a fog, Silver Fox and Coyote prepare the earth for the first people, vanishing before the people arrive.

The folklore of many cultures speaks of the fox's ability to shape-shift into a human being, usually an attractive young woman. For example, the tale of the "mysterious house-keeper" is found with slight variation among the North American Indians, the Inuit of Greenland and Labrador, and the Koriak of northeastern Siberia. The story tells of a hunter living alone who comes home at night to find his house clean and his supper cooking. Eventually he finds out that every morning a vixen comes to his home, takes off her skin and becomes a beautiful woman. He marries her and all goes well until he complains of a musky odour in the house. Her feelings hurt, the wife puts on the fox skin, turns back into a fox and runs away.

Other tales of transforming foxes, particularly those from Asia, portray a darker, more frightening creature. Despite the fox's role as divine messenger in Japan, Japanese tales also tell of evil foxes that haunt and possess people. In China, foxes are demons who become beautiful young men and women that seduce members of the opposite sex and then slowly consume their victim's being to prolong their own lives. The Chinese believed that the foxes lived through a succession of victims and could survive for eight hundred or a thousand years.

Reynard the fox is a major character in the folklore of medieval Europe. His exploits, perhaps based on misinterpreted observations of fox behaviour, include theft, fraud, rape and murder. At the end of his crime spree, Reynard goes unpunished. Instead, because of the cleverness he has shown, Reynard is named second in command by Noble, the lion king.

As these stories show, a mixture of respect and of fear and mistrust has been characteristic of the relationship between humans and foxes throughout history. Add to the folklore the hunter's search for sport, the trapper's desire for gain and the farmer's fears for livestock, and one can begin to understand the long-standing campaign of persecution waged against the fox.

Quarry, furbearer or vermin—one or more of these three predominant views of the fox can be found wherever foxes live, regardless of the species. Foxes are also well known as carriers of rabies, a disease fatal to both animals and humans, and as such have been the object of much attention in both North America and Europe. Our attempts to exploit or exterminate foxes have been many and varied, and often have had consequences far beyond what was imagined or intended—giving us reason to reflect on our past and future relationships with foxes and with animals in general.

FACING PAGE: *In this pose, the red fox looks every bit the crafty, cunning animal that folklore and myth portray.* ALAN & SANDY CAREY

72 FOXES

FOXES AS QUARRY

In a historical overview of the sport, David Macdonald notes that the earliest specific reference to fox hunting dates back to the fourth century B.C., but foxes were quarry long before that. Two thousand years ago, Alexander the Great was hunting foxes for sport and a seal from about that time shows a Persian horseman preparing to spear a fox. The Romans were pursuing foxes by A.D. 80. For centuries, the fox was considered a second-class quarry by western European hunters, but by 1420, when Edward, Second Duke of York, published *The Master of Game*, it was moving up in the ranks of animals worthy of chase.

Fox hunting grew enormously popular, particularly in Britain, where the sport developed into a social force affecting the laws and the landscape. While foxes were still minor game, Henry VIII had large tracts of land set aside in London so that he could hunt near his home. By 1750, the power of the hunting set was evident when the new Westminster Bridge opened. Macdonald relates that the second Duke of Bolton, weary of hacking the long way around via London Bridge from his home north of the Thames to his southern hunting grounds, led the parliamentary lobbying that resulted in the building of the new bridge. Land-use laws were passed to create land favourable for fox hunting and to preserve fox habitat, and hunt clubs paid farmers to spare breeding dens.

At the same time, the in-crowd in Germany was enjoying a variation of fox hunting—fox tossing. A lady and a gentleman stretched a narrow sling of webbing between them, and a fox was made to run across it. One can imagine the laughter of the fashionably dressed gathering as the fox was tossed into the air. A "good" toss could be over 7 m (23 feet) high. Augustus the Strong of Saxony was very fond of fox tossing and reportedly killed 687 foxes in one session.

Homesick colonists soon began to carry the British fox-hunting tradition around the world. They not only took their horses and hounds with them to the colonies but transported foxes as well. Hunters living in America's southern colonies thought that the native gray foxes provided poor sport, since they chose to climb trees rather than offer a proper chase. Red foxes were shipped from Britain and released in the colonies, where they soon began to breed with the existing northern population of native red foxes, expanding the red fox range and pushing out the gray foxes.

Australia had no foxes until the first ones arrived by ship in 1845, imported specifically for hunting. The species didn't really get a foothold until the 1870s, but by 1893, red foxes had colonized this new continent so well that Australia's first fox bounty scheme was established. The success of the fox came at a high price for Australia's native fauna, which were ill equipped to deal with this alien predator. Red foxes are at least partly to blame for the extinction of twenty species of local marsupials and are threatening many more today. After years of fruit-

FACING PAGE: *British colonists took their fox-hunting tradition with them to many places, including southern Saskatchewan, Canada. The settlers at Cannington Manor imported fox hounds from Iowa and dressed for hunts in full regalia. They also held hunt balls in evening dress, and it was often possible to see riders galloping over the prairie with mackintoshes draped over stiff-fronted shirts and dress trousers neatly tucked into tall boots.* SASKATCHEWAN ARCHIVES

less control programs, Australian scientists at the Cooperative Research Centre for Biological Control of Vertebrate Pest Populations are now working on a genetically engineered virus that will sterilize the foxes by tricking the vixens' immune systems into attacking male sperm. New Zealand escaped similar problems because of an 1867 act that prohibited importation of foxes. Tasmania had a close call in 1890 when two red foxes were imported for hunting; the foxes were later destroyed, however, and Tasmania has remained foxless.

Today, scarlet-coated riders still chase the red fox, following a tradition that is centuries old. The debate about fox hunting is heated. David Macdonald writes in *Running with the Fox*:

> Of course, the foxhunting debate defies consensus since different people put different values on each of the relevant factors. There is no common currency with which to equate units of suffering versus units of hedgerow versus units of employment versus units of cultural heritage and rural infra-structure and so forth. Scientists and economists can provide data to clarify the issues, but ultimately decisions will rest on values which are beyond the scope of science.

Many of the values that decide these issues are tied to our emotional reactions to foxes. These can be extreme. British author Michael Chambers has very definite views on the way people treat foxes:

> Foxes are, in a country [Britain] that styles itself "animal loving," subjected to a variety of very unpleasant practices: they are hunted, shot, . . . and baited with terriers, often being held, help-less, while the dogs savage them. I have heard reports of gratuitous cruelties so appalling as to challenge the reasonable imagination. This creature, beautiful, interesting and largely beneficial to man, is perhaps more than any other the plaything of the sadist and the victim of tradition.

Others have reactions that are more contradictory. For instance, Albert Pulling, who worked as a federal forester, game biologist and teacher during his eighty-odd years, can see the red fox both as a beautiful, intelligent creature and as a convenient target:

> The first fully grown red fox I ever shot came by when I was watching for squirrels. I was loaded with No. 6 shot, but he ran within 25 yards and I got off a round out of the full-choked barrel. I was about thirteen. . . . If you hunt in fox country—especially red fox terri-tory—you should enjoy some association with these beautiful and really foxy creatures.

The debate about hunting and shooting foxes will rage on, although I fail to understand the fun in chasing down, and sometimes killing, an exhausted fox, and it disturbs me to think that people can regard living creatures as mere moving targets.

The chilla lives on the plains, on the pampas and in the mountains of western South America, from Ecuador south through Peru and Chile. It was introduced to Tierra del Fuego in 1950 to control the rabbit population. Chillas have been heavily hunted for their fur and as pests and are now scarce throughout much of their range.

JEFF FOOTT

FOXES AS FURBEARERS

Sport hunters are only one group after the fox's skin. That skin—or more precisely, that marvellous fur—is the coveted prize of trappers around the world. The human quest for profit helped open up the North American continent and drastically altered the economic and social structures of its Native people. Peter C. Newman, in *Company of Adventurers: The Merchant Princes*, describes the chaos caused to Inuit culture by the arrival of the Hudson's Bay Company:

> Before the white man's arrival, the Inuit had little use for [arctic] fox pelts because they are too flimsy for clothing. Their only application was as a hand or face wipe—a kind of furry Kleenex—or for trimmings on children's clothes. The foxes' meagre back legs, their meatiest part, provided little nourishment.
>
> Since the Inuit had become dependent on the white man's goods, the fox trade became essential. But few realized how fundamentally the switch from hunting to trapping would disrupt their traditional society. . . . this shift from the primary role of hunter to trapper involved a radical switch in the aboriginals' sense of self-worth. Unlike trapping, hunting—especially of seal and polar bear—was a test of manhood, a dignified and courageous occupation in which each family celebrated the day's bounty. . . . The fox had to be trapped in winter when fur was at its prime, which meant abandoning most of the seal and caribou hunts. That in turn meant not only having to change social customs but also having to buy from the HBC the clothing and tools previously obtained as byproducts of seal and caribou kills.

In their book *Animal Welfare and Human Values*, Rod Preece and Lorna Chamberlain argue that fur trading severely undermined Native spiritual values as well:

> Traditional native culture was based on subsistence patterns of hunting and fishing, which were sufficient for food and raiment. With the arrival of the fur traders, this pattern was changed, and animals were hunted for furs to trade. No longer was the whole animal used. Indians' reverence for the animals they hunted was diminished by the commercial practice in which they now engaged.

Historically, the slaughter of foxes for their fur has provided an income for a great many people, and supplying the world demand for fox fur continues to be a large-scale operation.

Red fox furs, both red and silver phase, hang in bundles at a trading post, as they must have many years ago when the trade in furs was a driving force in opening up the North American continent.
GLEN & REBECCA GRAMBO

Doomed by its beauty, this intelligent creature will eventually be reduced to just one of a stack of pelts unless it escapes by chewing off its own foot. In that case, starvation or infection will finish what the trap began. WAYNE LYNCH

Between 1821 and 1891, the average number of red foxes killed annually for fur in North America was about 74,000—a total for seventy years of over five million foxes. In the former Soviet Union, more than 17 million red foxes died during the thirty-five years from 1924 to 1958. About 370,000 gray foxes were killed in the United States during the 1979–80 trapping season, and E. K. Fritzell states that as much as half the gray fox population of Wisconsin is "harvested" each year. Throughout the world, an average of 100,000 arctic foxes are killed each year, and, according to R. A. Garrott and L. E. Eberhardt, this has been the case for over one hundred years—at least 10 million arctic foxes dead in that time. These numbers count only foxes taken for their fur, ignoring the thousands killed for sport and as pests. Looking at the numbers, one is tempted to ask why there are any foxes left.

The answer lies in the fact that some fox species are incredibly resilient to this kind of pressure. Fewer foxes means more food and territory available for the survivors. Red foxes, gray foxes and arctic foxes all respond to increased mortality by increasing the size of their litters. Thus, a correlation can be drawn between fur price and litter size, since the foxes try to respond to increased trapping pressure when prices are good. Of course, there are limits to this capacity, and not all fox species are so tenacious.

For example, although the swift fox was never one of the major species hunted for its fur, its extirpation in Canada, and its dramatic decrease in population elsewhere, was due in part to trapping pressure. Between 1853 and 1877, Hudson's Bay Company records indicate an average of 4876 swift foxes taken per year (117,025 pelts total), even though the fur is not considered first quality. More recently, M. A. Mares and D. J. Schmidly state that the South American chilla (*Pseudalopex griseus*) has been hunted for its fur with such ferocity since the 1970s that populations are estimated to be only one-fifth of pre-1970 numbers. David Macdonald and Geoff Carr report that in 1975 Argentina exported about 200,000 chilla pelts; in 1978 the number had more than quadrupled, with nearly a million chilla pelts leaving the country. This species is now protected by law in Argentina, but there is so little enforcement that the fox is in danger of disappearing from the country. Argentina has been a huge exporter of fox pelts compared with its neighbour, Chile. From 1975 to 1979, Chile exported 1746 fox pelts. For the same period, Argentina exported 3,612,459 pelts—that works out to killing almost 1.3 foxes per square kilometre (approximately 3.5 foxes per square mile)! The chilla, the pampas fox (*Pseudalopex gymnocercus*) and the culpeo (*Pseudalopex culpaeus*) are all given an Appendix II listing by the Convention on International Trade in Endangered Species (CITES), which means that they may become endangered if trade in their pelts is not controlled.

Trapping wild foxes is a labour-intensive method of gathering fur that is subject to the vagaries of nature. The trapper must go where the foxes are and hope for a bountiful year. How much more efficient it would be to put the foxes in one accessible place where they

FACING PAGE: *This modern Norwegian fur farm is a far cry from the days of dumping foxes onto islands, but the objective remains the same: produce high-quality fox furs in the most economical way.* TOMMIE JACOBSSON/ NATURFOTOGRAFERNA

could easily be caught and killed. Modern fur farms hold foxes in cages where breeding, feeding and killing can all be done on schedule; however, fox fur farms certainly didn't begin that way.

During the 1830s, the Russian-American Company began dumping foxes onto various islands off Alaska. As a money-making scheme it was hard to beat. These islands, which had previously been fox-free, had large nesting seabird populations that provided a ready supply of fox food at no charge. The entrepreneurs let the foxes fend for themselves, returning every year to kill some. When the price for fox furs hit its peak in the 1920s and 1930s, red and arctic foxes were set loose on virtually every island from the Aleutian Islands to the Alexander Archipelago. When the two species were placed on the same island, the fur farmers soon discovered that the red fox eliminated the arctic fox, reducing profits. The death knell for the island "farms" was sounded in the 1930s and 1940s by the Great Depression and World War II, which combined to drop the bottom out of the fur market.

The tunnel vision of those involved in the island fur farms is responsible for a sad and lasting legacy. Whole breeding populations of seabirds were wiped out by the introduced foxes. The Aleutian Canada goose was driven to the brink of extinction by fox predation, and its breeding range today is confined to one small island. Christopher Lever wrote in 1987 that a survey of more than one hundred fox-filled islands south of the Alaskan peninsula showed a complete absence of nocturnal shorebirds.

Fur farming is no longer carried out on isolated islands. Instead, foxes are raised on farms or "ranches" in cages. The fur industry has fought a large-scale public relations battle with opponents of trapping and now offers garments made from ranch animals as a less cruel option. About 24 per cent of the fur coats manufactured in Canada and the United States are made from farm-produced skins.

Fur farms, like other businesses, are for-profit ventures; therefore, methods that maximize production and minimize cost are used. To economize on space, red and arctic foxes are often kept in cages less than a metre (a yard) square, with one to four animals to a cage. In such close quarters, foxes sometimes cannibalize each other. After about nine months, the foxes are killed using methods that won't damage their valuable fur. Commonly, two workers hold the animal down, and one clips an electric plate to its lip while the other inserts a probe into the fox's anus. The animal is then electrocuted by a current passed through its body. The fox may also be killed with strychnine or gassed by the exhaust from a vehicle. Whether fur ranching is more humane than trapping is clearly debatable.

Humans have learned to improve profits by raising foxes in cages rather than on islands, but we seem to have missed a more obvious lesson: closed ecosystems are especially vulnerable when alien animals are introduced. Fur farming of silver-phase red foxes is increasing in Iceland, running the grave risk of introducing a non-native predator to a country where humans have tried for the last seven hundred years to eliminate the only native terrestrial mammal—the arctic fox—because of its predatory nature. Any poultry farmer can tell you the difficulty of building an enclosure that is impenetrable to foxes. What happens to the native animals, particularly the arctic fox, if enough red foxes escape to form a breeding population? Who will explain to farmers that they are now faced with a larger, more aggressive predator?

Peeking cautiously around the rocks of its island home, this arctic fox shows the blue-phase colouring found more often in coastal areas, where there is less snow accumulation, giving the more typical white-phase fox less of a camouflage advantage. ART WOLFE

88 FOXES

FOXES AS VERMIN

The fox's long-standing reputation as a predator of domestic stock makes it a welcome sight in the cross hairs of many a farmer's rifle scope. If it is not after the chickens or sheep, the fox could be taking game reserved for the hunter's table. But how much of the fox's infamy is deserved?

In the first place, any farmer raising grain should be putting out a welcome mat for foxes. Foxes eat lots of grain-nibbling rodents—zoologist Bernhard Grzimek reports that one red fox was found with forty-eight voles in its stomach! However, the grain farmer's neighbours who raise sheep may object to foxes in their pastures.

The livestock farmer sometimes convicts the fox on circumstantial evidence. Foxes seen near lambing or calving areas may appear to be hunting live prey, when they are usually searching for placentas and stillborn animals. Chicken bones or wool found near a fox den, like the deer bones that may also be found there, are often remnants from scavenged carrion rather than the remains of animals the foxes have killed themselves. Leonard Lee Rue III quotes Paul Errington: "The cured ham that some Iowa hunters retrieved from a fox den does not indicate that the fox killed a pig."

Although studies indicate that domestic stock is a minor item on most fox menus, foxes undeniably kill poultry and have been known to take lambs. Whether this is the result of the fox's determination or the farmer's livestock management practices is another question. Chickens loose in the farmyard or lambs weakened from rough spring weather are a great temptation to a hungry fox. Given the choice of a hard night's hunt or a relatively easy meal, the opportunistic fox takes the obvious path. Since the loss of even one lamb or a few chickens can seem like a lot to a subsistence farmer counting on every piece of livestock, the fox remains an unwelcome visitor.

Different people cope with foxes in different ways. One farm family that I know raises extra chickens each year to compensate for those the local foxes will take. They enjoy watching the foxes all summer and don't want to shoot them, so they have adopted this arrangement. Another family living a few kilometres away also enjoys watching the foxes that den by their house every year. And every year, they shoot three or four of those foxes to protect the poultry in their yard. They say they don't like doing it, but foxes and poultry don't mix; this is their solution.

While farmers are concerned about their livestock, hunters and gamekeepers regard the fox as an unwanted competitor for game. In South Dakota, where I grew up, the ring-necked pheasant is an important natural resource that brings hunters and money into the

FACING PAGE: *These young foxes seem content at their less-than-secluded den. Red foxes adapt better than many other animals to human presence—a red fox in Oxford, England, even took up residence amid the bustle and noise of a busy factory warehouse.*
DANIEL J. COX

state. During the 1970s, a group called Pheasants Unlimited reportedly killed over 150,000 foxes and encouraged 4-H clubs to earn points in a "Pheasant Restoration Contest" by joining in the killing. Such systematic slaughter is far from uncommon, although the effectiveness is questionable, since fox populations recover rapidly and vacant territories are quickly filled by other foxes.

Because of complicating factors such as bird losses due to disease or bad weather, it is not clear how the presence of foxes affects game bird populations. Birds are not a staple item in the fox diet and are taken only when they are particularly vulnerable or when a fox's usual prey is scarce. There is no doubt that when ground-nesting birds and their eggs are abundant, especially if they are concentrated in small areas, foxes seize their opportunity and can take a heavy toll. A. B. Sargeant and others have estimated that red foxes took more than 800,000 ducks annually from the prairie pothole region of North America during spring and early summer from 1969 through 1973. Some investigators considered red foxes the main predator of ducks and duck eggs in the region. Eggs seem to be a particular delicacy, and foxes often cache a great number for future consumption. At a local level, foxes may therefore cause a marked reduction in the number of game birds available to the hunter's gun.

The general conclusion of several other studies, however, has been that foxes have little effect on game bird populations on a large scale. For example, when surveys of ruffed grouse nests in New York State showed that 39 per cent of the nests were destroyed by predators, a large portion of the loss was attributed to red foxes; however, a predator-control program failed to bring about a rise in grouse numbers. In another example, Leonard Lee Rue III tells of the ring-necked pheasant population on an island in Lake Michigan that fell at the same rate as that on the mainland, despite the fact that there were no foxes on the island. More research is needed to discover the role that foxes, and other predators, play in controlling game bird numbers.

Even in Iceland, where fear for the valuable eider colonies has helped make the arctic fox public enemy number one, research has shown that if it weren't for the foxes, the down business might not be as productive. Researchers Päll Hersteinsson and Anders Angerbjörn describe how when arctic foxes disappeared from part of western Iceland at the turn of the century, down harvests dropped by 66 to 75 per cent. The eiders dispersed over a larger area during the mating season, perhaps because of reduced fear of foxes. For the down gatherers, this meant that finding the nests took longer and was less profitable. The intricate relationships between living things rarely lend themselves to simple models of analysis and management.

The arctic fox population in Scandinavia shows the unintentional results of a typical attempt to simplify nature. The once-abundant arctic fox is now classified as endangered in mainland Norway, Sweden and Finland, and shows few signs of recovery. In fact, there are almost six times as many arctic foxes in Iceland, where they are shot on sight, as in all of Scandinavia, where they have been protected for over half a century. Although the initial decline was probably due partly to heavy trapping, Hersteinsson and Angerbjörn believe the real problem was caused by an extermination campaign directed against another predator—the wolf.

Removing wolves had two serious consequences for the arctic fox. Wolf kills that had provided a major source of food for the foxes were greatly reduced, and the lack of wolves cleared the way for the red fox. The red fox is much less tolerant of the arctic fox than the wolf is and can push out its smaller relative. Red foxes invaded arctic fox denning areas, preying on their cubs and spreading diseases among the arctic foxes. Whether anything can be done to help the Scandinavian arctic fox population recover is unknown. Drastic reduction of the arctic fox population was not a planned outcome of the wolf-extermination campaign but it happened anyway—because of a lack of understanding. We are slowly learning that exterminating a species to solve a problem rarely works and often creates a new set of problems.

LEFT: *The stereotypical red fox in pursuit of a poultry dinner, this vulpine acrobat was later joined by a less daring companion on the ground.* ALAN & SANDY CAREY

ABOVE: *Foxes may overcome their fear of humans, as this visitor to a European airport shows. This phenomenon causes some concern for health experts charged with controlling the spread of rabies, but the possibility of direct contact is less alarming than the tendency for wild fox populations to act as reservoirs of the disease.* HANS RING/NATURFOTOGRAFERNA

FOXES AND RABIES

Some of the most determined extermination campaigns in history have been waged against the red fox. The reason for this massive slaughter is that wild red fox populations, along with gray and arctic foxes, serve as a reservoir for rabies. In the Northern Hemisphere, red foxes are the chief carriers of the disease and the chief victims. Rabies, a disease that affects the central nervous system, is one of the oldest recognized diseases and one of the most feared. Unless immediately treated with a series of immunizations, rabies victims die in agony; every year almost twenty-five thousand people perish this way. Around the world, hundreds of thousands of animals have been killed and millions of taxpayer dollars spent in an attempt to eradicate this disease. The scale of the battle is hard to comprehend.

In 1952, for example, a control program in the province of Alberta, Canada, had 180 trappers each working about 50 km (30 miles) of trapline. They were issued 6000 cyanide capsules and 429,000 strychnine cubes. The minimum kill for an eighteen-month period was 50,000 foxes, 35,000 coyotes, 43,000 wolves, 7500 lynxes, 1850 bears, 500 skunks, 64 cougars, 1 wolverine and 4 badgers. At the same time, farmers supporting the program were given 75,000 cyanide shells and 163,000 strychnine pellets, resulting in the deaths of another 60,000 to 80,000 coyotes.

The most obvious result of this program was a huge jump in the number of deer and moose because there were no longer enough predators to keep a check on the populations. So many grazers put a huge strain on the rangelands, causing long-term damage and reducing their ability to support coveted big game. This damage lasted much longer than the check on rabies, which soon spread again as carrier populations recovered.

In Europe, intense trapping, shooting and den-gassing campaigns have been the rule for years. In 1975, it was estimated that 180,000 red foxes were being killed each year in West Germany alone. Government hunters systematically carried out a program of extermination as countries tried frantically to stop the spread of the disease. Did the killing stop the spread of rabies? With one exception—Denmark—the answer is no.

The attempt to control rabies by severely reducing the red fox population failed for a number of reasons. The first is the resiliency of the species. The foxes were capable of sustaining huge losses and still maintaining viable populations. David Macdonald and Philip Bacon state that the elimination of individuals may have actually increased the spread of rabies by breaking up the established social order among fox groups and causing more hostile encounters in which biting infected new foxes. There had to be a better method for controlling the disease.

In 1978, a hiker walking through the Rhône River valley in the Swiss Alps might have been surprised to stumble over a chicken head. Chicken heads, laced with live rabies vaccine, had been spread across the valley in an attempt to halt an oncoming tide of rabies. Scientists hoped that foxes would eat the bait, thus immunizing themselves and forming a living barrier against the disease. The plan worked, and hope for a new solution to the rabies problem grew.

Chicken heads are messy and hard to get in massive numbers, so scientists developed a sort of fox dessert wafer made of fishmeal, bonemeal and fat, with an imbedded plastic packet of rabies vaccine. The wafer was a hit with foxes, and since 1983 over 5.2 million baits have been distributed. Over 70 per cent of foxes in the target areas take the bait and become immunized. This means that there are no longer enough susceptible animals in these areas to continue the chain of transmission.

Today Switzerland is virtually rabies-free except for isolated cases near its border regions, and the occurrence of rabies is dropping rapidly in other countries making extensive use of oral vaccines. In Canada, the province of Ontario is reporting a dramatic drop in fox rabies following an aerial bait-vaccination campaign. For the first time in human history, we are working towards eliminating a disease found in free-ranging animals using a solution other than wholesale slaughter.

A LOOK TO THE FUTURE

The cost/benefit ratio of the relationship between humans and foxes is heavily weighted against the fox. Human activity is sometimes helpful to foxes. For example, increased edge habitats created by clearing woodlands provide better hunting for red and gray foxes. In almost every other case, however, human interaction with foxes is directly and deliberately detrimental to the fox.

Michael Chambers writes, "Man will massacre his fellow creatures for idle pleasure to see them fall; for profit often to satisfy the utterly frivolous vanity of his fellow man." Red, gray and arctic foxes are rarely killed for food but are slaughtered in huge numbers for entertainment or for fur. These foxes are able to withstand the high mortality rate and are not endangered species. Is this sufficient justification to allow the killing to continue unquestioned?

If we fall into the trap of equating population numbers with a species' well-being, we become concerned about animals only when their situation is growing desperate, and we value the preservation of a species while neglecting the welfare of the individual. Does the fact that there are plenty of red foxes make it more acceptable for a family of kits to die slowly of starvation because a hunter shot their mother?

Dying is part of the cycle of life—a fact often used by humans to justify their taking of animal lives. Animals unquestionably die in the wild, sometimes horribly, killed by other animals or by the forces of nature. Nature has no moral sense, however, and animals make no boasts about their ethics. A fox need feel no pity when it kills a vole, and nature cannot be rebuked for a lack of compassion if the same fox dies of disease.

Humans cannot claim these exemptions, and when a human takes the life of another animal it should be done because of a true need, not to satisfy a frivolous whim. We humans pride ourselves on our ability to think and reason, believing that this sets us apart from the other animals, yet all too often we behave unthinkingly and act unreasonably in our dealings with them. It all comes down to what type of values we place on wildlife.

Many people would say that a fox's only worth is available on its death, when it becomes a trophy of sport or a pelt for sale, but living foxes have a very practical value. According to Emily and Per Ola D'Aulaire, following a systematic removal of foxes from Door County, Wisconsin, the mouse population rose dramatically, necessitating restoration of the predators. By consuming large numbers of rodents, foxes and other hunters of small mammals reduce crop losses and play a part in controlling rodent-borne diseases. They also perform a valuable clean-up service when they scavenge carrion. As researcher H. J. Egoscue states, fox

FACING PAGE: *A morning spent watching these kits at play may have no monetary value, but in the joy given to the viewer, it is beyond price.* ALAN & SANDY CAREY

dens, such as the large complexes of the swift fox, are mini-ecosystems that provide food, breeding space and safety for many kinds of animals, including burrowing owls, spiders, centipedes and beetles. Arctic fox dens perform a similar function for plants, encouraging the growth of important fodder grasses that are often absent on the surrounding tundra. Like all living things, foxes are part of an intricately balanced scheme that doesn't hold up well under sledgehammer tactics.

Foxes have another value, one that is more difficult to describe or quantify. It comes from seeing such a lovely creature, so perfectly fitted for the life it leads. What value can be placed on those few privileged moments when we are allowed to be part of a fox's world? How does one put a price on the pleasure of watching fox kits romping in the sunshine? Leonard Lee Rue III writes:

> The aesthetic value to the average person who gets a glimpse of a red fox daintily going about its foxy business is beyond calculation. The red fox is neither good nor bad; it is merely a red fox, admirably fulfilling the niche for which it was created.

Foxes are creatures of the shadows, feared and hunted through history. They are quarry to be chased, fur to be sold or a threat to be removed. They are loving parents, faithful partners and hunters of unmatched skill. Foxes are all of these things and so much that still remains a mystery. But there is no mystery about the thrill that I feel when I am lucky enough to see a fox, flickering like a flame through a snowy forest. May we never take such magic for granted.

FACING PAGE: GLEN &
REBECCA GRAMBO

Appendix # FOX SPECIES OF THE WORLD

There is disagreement on the taxonomy of the foxes. I have chosen to use the classification summarized by J. W. Sheldon in *Wild Dogs: The Natural History of the Nondomestic Canidae*. Distribution data are taken from the same source.

SPECIES	COMMON NAME(S)	DISTRIBUTION/COMMENTS
Alopex lagopus	Arctic fox	Occupies the Arctic or tundra regions of North America, Eurasia, Scandinavia, Spitsbergen, Greenland, Iceland and many islands of the Arctic, North Atlantic and North Pacific oceans.
Atelocynus microtis	Small-eared dog Small-eared zorro (fox) Zorro negro	Found in the Amazon, Upper Panama and Orinoco basins in Colombia, Ecuador, Brazil, Peru and Venezuela. Prefers tropical forest habitat from sea level to 1000 m (3000 feet). Only canid to occupy this type of habitat.
Cerdocyon thous	Crab-eating fox	Found across the Brazilian subregion, except for the Amazon Basin Lowlands, and from Colombia and Venezuela southward through Brazil into northern Argentina and Uruguay.
Fennecus zerda	Fennec fox	Occupies the northernmost African countries, from Morocco through Algeria, Tunisia, Niger, Libya, Egypt and the Sudan. Prefers desert and subdesert habitats. CITES Appendix II listing.
Otocyon megalotis	Bat-eared fox	Two populations: one is found from Botswana, southern Angola and western Zambia southward into South Africa; the second group is found in Somalia, Ethiopia and southern Sudan and extending southward to Tanzania. Prefers arid or semi-arid environments.

Pseudalopex culpaeus	Culpeo (fox) Andean wolf Coloured fox Colpeo or culpaeo fox Large fox	Occupies western coastal region of South America from southern Colombia and Ecuador southward through Peru, western Bolivia, Chile and Argentina, to Tierra del Fuego. Prefers arid to semi-arid environments in a wide range of habitats. CITES Appendix II listing.
Pseudalopex griseus	Chilla Argentine (gray) fox Chico gray fox Pampa fox Little gray fox	Found in Chile and Argentina below 25° south latitude; in the plains and low mountains of Chile, Argentina and Patagonia. Introduced to Tierra del Fuego in 1950. CITES Appendix II listing.
Pseudalopex gymnocercus	Pampas (gray) fox Azara's fox Paraguayan fox	Found in east-central South America from southeastern Brazil to Paraguay, Uruguay and northeastern Argentina south to the Rio Negro. Habitat destruction has removed these foxes from portions of their original range. CITES Appendix II listing.
Pseudalopex sechurae	Sechura(n) fox Peruvian desert fox Sechuran desert fox	Found in a small portion of the coastal zones of northwestern Peru, including the Sechuran desert and southwestern Ecuador. Prefers arid habitats.
Pseudalopex vetulus	Hoary fox Field fox Small-toothed dog	Found in the central portions of Brazil in the states of Mato Grosso, Goiás, Minas Gerais and São Paulo. Prefers open country.
Urocyon cinereoargenteus	Gray fox	Northward limit of range found along U.S./Canadian border to the east and west of Great Lakes. Southward throughout North America except for the northern Rocky Mountain region, the northern end of the Great Basin and Washington State. South through Central America and into Venezuela and Colombia.
Urocyon littoralis	Island gray fox	Found on six of eight of the Channel Islands off the southern California coastline. Listed by the state of California as "threatened." Slightly smaller and with two fewer tail vertebrae than *Urocyon cinereoargenteus*.
Vulpes bengalensis	Bengal fox	Found throughout Indian subcontinent from southernmost areas up into Nepal and the Indian state of Assam on the east and Pakistan on the west. Found in the Himalayan foothills up to 1500 m (4500 feet). Prefers scrubby and open habitats.
Vulpes cana	Blanford's fox Hoary fox Afghan fox Baluchistan fox King fox	Range is not well documented. From northeastern Iran to Afghanistan and northwestern Pakistan. In Israel and the Sinai; two specimens reported from Oman. Prefers mountain-steppe habitat. CITES Appendix II listing.

Vulpes chama	Cape fox Silver jackal Silver fox	Originally found throughout the arid and semi-arid western areas of southern Africa. Current range includes northern Cape Province, southern and central Namibia, Botswana, southwestern Angola, Zimbabwe and the Transvaal. Prefers arid environments; never reported in forested areas.
Vulpes corsac	Corsac fox	Widely distributed in Asia; from the Azov Sea west through China and Mongolia to the Transbaikalian steppes and into northern Manchuria. Some found in northeastern Iran and northern Afghanistan. Prefers steppes and subdesert zones.
Vulpes ferrilata	Tibetan sand fox	Found in Tibet and in the Mustang District of northern Nepal. Prefers plateaus in alpine desert habitat at or above 3000 m (9800 feet).
Vulpes macrotis	Kit fox	Occupies arid and semi-arid regions of the western U.S. and northern Mexico, including the Baja Peninsula. U.S. range includes extreme southwestern corner of Oregon and extends southward to parts of Idaho, Nevada, Utah, Arizona and New Mexico.
Vulpes pallida	Pale fox Pallid fox Sand fox	Range cuts a broad path across the Saharan and sub-Saharan regions of Africa from Senegal and Mauritania on the west coast through Mali, Niger, Nigeria, northern Cameroon, Chad and the northern portions of the Sudan.
Vulpes rüppelli	Rüppell's fox Sand fox	Occupies arid regions of northern Africa, the Arabian Peninsula and western Asia. Extremely well adapted to desert life.
Vulpes velox	Swift fox	Original range included much of the plains of west-central North America from Texas north to the prairies of southern Canada. This range is now restricted and re-introduction programs are being carried out in Canada. Prefers open plains, short- and medium-grass prairies.
Vulpes vulpes	Red fox	Largest natural distribution of any living terrestrial mammal except humans. Occupies most of the Northern Hemisphere above 30° north latitude: all of Asia except for extreme southeastern portion, throughout Europe, northern Africa, and North America as far south as central Texas. Introduced to Australia and some Pacific islands. Tundra marks northern edge of range.

BIBLIOGRAPHY

The following sources were the most important in the compilation of this book.

Ables, E. D. 1975. Ecology of the red fox in North America. In *The wild canids: Their systematics, behavioral ecology, and evolution.* Ed. M. W. Fox, 216–36. New York: Van Nostrand Reinhold.

Bacon, P. J., and D. W. Macdonald. 1980. To control rabies: Vaccinate foxes. *New Sci.* 87:640–45.

Barden, R. 1993. Foxes. In *Endangered wildlife of the world,* 442–47. North Bellmore, NY: Marshall Cavendish.

Brechtel, S. H., L. N. Carbyn, D. Hjertaas and C. Mamo. 1993. The swift fox reintroduction feasibility study: 1989 to 1992. Unpublished report to Western Wildlife Directors.

Cahalane, V. H. 1961. *Mammals of North America.* New York: Macmillan.

Carbyn, L. N., H. J. Armbruster and C. Mamo. 1994. The swift fox reintroduction program in Canada from 1983 to 1992. In *Restoration of endangered species: Conceptual issues, planning and Implementation.* Ed. M. L. Bowles and C. J. Whelan, 247–71. Cambridge: Cambridge University Press.

Chambers, M. 1990. *Free spirit.* London: Methuen London.

Chesemore, D. L. 1975. Ecology of the arctic fox (*Alopex lagopus*) in North America—a review. In *The wild canids: Their systematics, behavioral ecology, and evolution.* Ed. M. W. Fox, 143–63. New York: Van Nostrand Reinhold.

D'Aulaire, E., and P. O. D'Aulaire. 1980. The importance of being wily. *Natl. Wildl.* 18:24–28.

Durrell, G. 1961. *The whispering land.* London: Rupert Hart-Davis.

Egoscue, H. J. 1979. *Vulpes velox. Mamm. Spec.* No. 122, 1–5.

Ewer, R. F. 1973. *The carnivores.* Ithaca, NY: Cornell University Press.

Fox, M. W. 1975. *The wild canids: Their systematics, behavioral ecology, and evolution.* New York: Van Nostrand Reinhold.

Fritzell, E. K. 1987. Gray fox and island gray fox. In *Wild furbearer management and conservation in North America.* Ed. M. Novak, J. A. Baker, M. E. Obbard and B. Malloch, 408–21. North Bay, ON: Ontario Trappers Association.

Fritzell, E. K., and K. J. Haroldson. 1982. *Urocyon cinereoargenteus. Mamm. Spec.* No. 189, 1–8.

Garrott, R. A., and L. E. Eberhardt. 1987. Arctic fox. In *Wild furbearer management and conservation in North America.* Ed. M. Novak, J. A. Baker, M. E. Obbard and B. Malloch, 394–407. North Bay, ON: Ontario Trappers Association.

Grzimek, H. C. B. 1975. *Grzimek's animal life encyclopedia,* 12:243–54, 267–80. New York: Van Nostrand Reinhold.

Henry, J. D. 1986. *Red fox: The catlike canine.* Washington, DC: Smithsonian Institution Press.

Hersteinsson, P., and A. Angerbjörn. 1989. The arctic fox in Fennoscandia and Iceland: Management problems. *Biol. Conserv.* 49:67–81.

Iriarte, J. A., and F. M. Jaksić. 1986. The fur trade in Chile: An overview of seventy-five years of export data (1910–1984). *Biol. Conserv.* 38:243–53.

Kaplan, C., ed. 1977. *Rabies: The facts.* Oxford: Oxford University Press.

Macdonald, D. W. 1980. Social factors affecting reproduction amongst red foxes (*Vulpes vulpes* L., 1758). In *The red fox: A symposium on behaviour and ecology*, Biogeographica Vol.18. Ed. E. Zimen, 123–75. The Hague: Dr. W. Junk b.v. Publishers.

Macdonald, D. W. 1984. *The encyclopedia of mammals.* New York: Facts on File.

Macdonald, D. W. 1988. *Running with the fox.* New York: Facts on File.

Macdonald, D.W., and G. Carr. 1981. Foxes beware: You are back in fashion. *New Sci.* 89:9–11.

McGrew, J. C. 1979. *Vulpes macrotis. Mamm. Spec.* No. 123, 1–6.

MacInnes, C. D. 1987. Rabies. In *Wild furbearer management and conservation in North America.* Ed. M. Novak, J. A. Baker, M. E. Obbard and B. Malloch, 910–29. North Bay, ON: Ontario Trappers Association.

Mares, M. A., and D. J. Schmidly, eds. 1991. *Latin American mammalogy: History, biodiversity and conservation.* Norman, OK: Oklahoma Museum of Natural History.

Mercatante, A. S. 1988. *The Facts on File encyclopedia of world mythology and legend.* New York: Facts on File.

Morris, D. 1990. *Animalwatching.* London: Jonathan Cape.

Newman, P. C. 1991. *Company of adventurers.* Vol. 3, *Merchant princes.* Toronto: Penguin Group.

Novak, M., J. A. Baker, M. E. Obbard and B. Malloch, eds. 1987. *Wild furbearer management and conservation in North America.* North Bay, ON: Ontario Trappers Association.

O'Farrell, T. P. 1987. Kit fox. In *Wild furbearer management and conservation in North America.* Ed. M. Novak, J. A. Baker, M. E. Obbard and B. Malloch, 422–31. North Bay, ON: Ontario Trappers Association.

People for the Ethical Treatment of Animals. Ranch-raised fur: Captive cruelty. *PETA Factsheet* Wildlife #3. Washington, DC: People for the Ethical Treatment of Animals.

Preece, R., and L. Chamberlain. 1993. *Animal welfare and human values.* Waterloo, ON: Wilfrid Laurier University Press.

Pulling, Albert Van S. 1973. *Game and the gunner.* New York: Winchester Press.

Rue, Leonard Lee, III. 1969. *The world of the red fox.* New York: J. B. Lippincott.

Sargeant, A. B., R. J. Greenwood, M. A. Sovada and T. L. Shaffer. 1993. Distribution and abundance of predators that affect duck production—prairie pothole region. U.S. Dept. Interior, Fish and Wildlife Service, Resource Publication 194.

Scott-Brown, J. M., S. Herrero and J. Reynolds. 1987. Swift fox. In *Wild furbearer management and conservation in North America.* Ed. M. Novak, J. A. Baker, M. E. Obbard and B. Malloch, 432–41. North Bay, ON: Ontario Trappers Association.

Seton, E. T. 1909. *Life-histories of northern animals.* New York: Arno Press.

Sheldon, J. W. 1992. *Wild dogs: The natural history of the nondomestic Canidae.* New York: Academic Press.

Sinclair, S. 1985. *How animals see.* New York: Facts on File.

Skrobov, V. D., and E. A. Shirokovskaya. 1968. The role of the arctic fox in improving the vegetation cover of the tundra. *Problems of the North,* No. 11, 123–28.

Smith, T. G. 1976. Predation of ringed seal pups (*Phoca hispida*) by the arctic fox (*Alopex lagopus*). *Can. J. Zool.* 54:1610–16.

Trapp, G. R., and D. L. Hallberg. 1975. Ecology of the gray fox (*Urocyon cinereoargenteus*): A review. In *The wild canids: Their systematics, behavioral ecology, and evolution.* Ed. M. W. Fox, 164–78. New York: Van Nostrand Reinhold.

Van der Wall, S. B. 1990. *Food hoarding in animals.* Chicago: University of Chicago Press.

Weintraub, P. 1993. Vaccines go wild. *Audubon* 95:16.

Winkler, W. G., and K. Bögel. 1992. Control of rabies in wildlife. *Sci. Am.* 266:86–92.

Zimen, E., ed. 1980. *The red fox: A symposium on behaviour and ecology,* Biogeographica 18:1–285. The Hague: Dr. W. Junk b.v. Publishers.

INDEX